THE
BREAKTHROUGH

Also by Charles Graeber

The Good Nurse: A True Story of Medicine,
Madness, and Murder

THE
BREAKTHROUGH

Immunotherapy and the
Race to Cure Cancer

~

CHARLES GRAEBER

TWELVE

NEW YORK   BOSTON

Twelve
Hachette Book Group
1290 Avenue of the Americas, New York, NY 10104

twelvebooks.com

twitter.com/twelvebooks

First Edition: November 2018

Twelve is an imprint of Grand Central Publishing. The Twelve name and logo
are trademarks of Hachette Book Group, Inc.

The publisher is not responsible for websites (or their content) that are not
owned by the publisher.

The Hachette Speakers Bureau provides a wide range of authors for
speaking events. To find out more, go to www.hachettespeakersbureau.com
or call (866) 376-6591.

Library of Congress Cataloging-in-Publication Data has been applied for.

ISBNs: 978-1-4555-6850-5 (hardcover), 978-1-4555-6849-9 (ebook)

Printed in the United States of America

LSC-C

10 9 8 7 6 5 4 3 2 1

For Diann Waterbury Graeber
My mother and our survivor

Contents

Preface

*It seemed to me then, and still does, that some such
built-in immunologic mechanism ought to exist for
natural defense against cancer in humans.*

—LEWIS THOMAS, 1982

Cancer is alive. It's a normal cell, mutated and changed, and it continues to change in the body.

Unfortunately, a cancer drug does not mutate or change.

A drug may poison or starve the cancer for a time, but whatever cancer cells remain will continue to mutate. It only takes one. The drug dances with cancer, but cancer dances away.

As a result, these types of drugs are unlikely to ever truly cure cancer.

But we have killers in our bodies, and scouts and soldiers, a dynamic network of cells more nimble than any cancer. This is our immune system, a living defense as old as life itself.

This system mutates. It adapts. It learns and remembers and matches an innovating disease step for step.

It's our best tool to cure cancer.

And we have finally discovered how to unleash it.

This is the breakthrough.

THE
BREAKTHROUGH

Introduction

The good physician treats the disease; the great physician treats the patient who has the disease.

—Sir William Osler, 1849–1919

Until very recently we've had three main methods for treating cancer. We've had surgery for at least three thousand years. We added radiation therapy in 1896.[1] Then in 1946, chemical warfare research led to the use of a mustard gas derivative to kill cancer cells. Those poisons were the foundation for chemotherapy.

These "cut, burn, and poison" techniques are currently estimated to be able to cure cancer in about half of the people who develop the disease. And that's remarkable, a true medical accomplishment. But that leaves the other half of cancer patients. Last year, in the United States alone, that translated to nearly six hundred thousand people who died of the disease.

The fight was never fair. We've been pitting simple drugs against creative mutating versions of our own cells, trying to kill the bad ones while sparing the good ones and making ourselves sick in the process. And we've been doing that for a very long time.

But now we have added a new and very different approach— one that doesn't act directly on cancer, but on the immune system.

~

Our immune system has evolved over 500 million years into a personalized and effective natural defense against disease. It is a complex biology with a seemingly simple mission: to find and destroy anything that's not supposed to be in our bodies. Cells of the immune system are on constant patrol, hundreds of millions of them circulating throughout the body, slipping in and out of organs, searching out and destroying invaders that make us sick and body cells that have become infected, mutated, or defective—cells like cancer.

Which raises the question: Why doesn't the immune system fight cancer already?

The answer is, it does, or tries to. But cancer uses tricks to hide from the immune system, shut down our defenses, and avoid the fight. We don't stand a chance, unless we change the rules.

Cancer immunotherapy is the approach that works to defeat the tricks, unmask cancer, unleash the immune system, and restart the battle. It differs fundamentally from the other approaches we have to cancer, because it does not act upon cancer at all, not directly. Instead it unlocks the killer cells in our own natural immune system and allows them to do the job they were made for.

~

Cancer is us. It's the mistake that works. Cells in the body regularly go rogue, their chromosomes knocked out by particles of sunlight or toxins, mutated by viruses or genetics, age, or sheer randomness. Most of these mutations are fatal to the cell, but a few survive and divide.

99.9999 percent of the time, the immune system successfully recognizes these mutant cells and kills them. The problem is that rogue 0.0001 percent cell, the one that the immune system doesn't recognize as an invader and does not kill. Instead, eventually, that 0.00001 percent cell kills us.[2]

~

Cancer is different. It does not announce itself like the flu or any other disease, or even a splinter. It doesn't seem to sound an alarm in the house of the body, or provoke an immune response, or show symptoms of immune battle: no fever or inflammation or swollen lymph glands, not even a sniffle. Instead, the tumor is suddenly discovered, an unwelcome guest that has been growing and spreading out, sometimes for years. Often by then it is too late.

To many cancer researchers, this apparent lack of immune response to cancer meant that the goal of *helping* an immune response to cancer was futile—because there was nothing to help. Cancer was assumed to be too much a part of our selves to be noticed as "non-self." The very concept of cancer immunotherapy seemed fundamentally flawed.

But throughout history, physicians had recorded rare cases of patients whose cancers apparently cured themselves. In a prescientific age these "spontaneous remissions" were seen as the work of magic or miracle; in fact, they are the work of an awakened immune system. For more than a hundred years researchers tried and failed to replicate those miracles through medicine, to vaccinate or spark an immune response to cancer similar to those against other formerly devastating diseases like polio, smallpox, or the flu. There were glimmers of hope, but no reliable treatments. By the year 2000, cancer immunologists had cured cancer in mice hundreds of times, but could not consistently translate those results to people. Most scientists believed they never would.

That changed radically and recently. Even for physicians, this change was invisible until it was at the doorstep. One of our best modern writers on the subject of cancer, Dr. Siddhartha Mukherjee, does not even mention cancer immunotherapy in his nonetheless excellent Pulitzer Prize–winning biography of the disease, *The Emperor of All Maladies*. That book was published in 2010, only five months before the first of the new-generation immunotherapeutic cancer drugs received FDA approval.

That first class of cancer immunotherapy drugs would be called "checkpoint inhibitors." They came from the breakthrough discovery of specific tricks, or "checkpoints," that cancer uses like a secret handshake, telling the immune system, *Don't attack.* The new drugs inhibited those checkpoints and blocked cancer's secret handshake.

In December 2015 the second of these checkpoint inhibitors[3] was used to unleash the immune system of former president Jimmy Carter. An aggressive cancer had spread through his body and he wasn't expected to survive; instead, his immune cells cleared the cancer from his liver and brain. The news of the ninety-one-year-old president's miraculous recovery[4] surprised everyone, including the former president himself. For many people, "that Jimmy Carter drug" was the first and only thing they'd heard about cancer immunotherapy.

But the breakthrough isn't any one treatment or drug; it's a series of scientific discoveries that have expanded our understanding of ourselves and this disease and redefined what is possible. It has changed options and outcomes for cancer patients, and opened the door to a rich and uncharted field of medical and scientific exploration.

These discoveries validated an approach to beating cancer that is conceptually different from the traditional options of cut, burn, or poison, an approach that treats the patient rather than the disease. For the first time in our age-old war with cancer, we understood what we were fighting, how cancer was cheating in that fight, and how we might finally win. Some call this our generation's moon shot. Even oncologists, a cautious bunch, are using the C word: *cure*.

Hype can be dangerous, just as false hope can be cruel. There's a natural tendency to invest too much hope in a new science, especially one that promises to turn the tables on a disease that has, in some way, touched every person's life. Nevertheless, these aren't overhyped theories or anecdotal wonder cures, but proven medicines based on solid data. Immunotherapy has gone from being a dream to a science.

Right now there are only handful of immunotherapies available. Less than half of all cancer patients have been shown to respond to these drugs. Many who do respond profoundly, with remissions measured not in extra weeks or months of life, but in lifetimes. Such transformative, durable responses are the unique promise of the cancer immunotherapeutic approach, and part of what makes it attractive to patients, but it's important to note that that promise is different from a guarantee for any one outcome for any individual patient. We still have work to do to widen the circle of responders and truly find a cure. But the door is open now, and we've only just begun.

Several of the immunotherapists I interviewed compared the discovery of these first cancer immunotherapy drugs to that of penicillin.[5] As a drug, penicillin immediately cut infection rates, cured some bacterial diseases, and saved millions of lives. But as a scientific breakthrough, it redefined the possible and opened a fertile new frontier for generations of pharmaceutical researchers. Nearly one hundred years after the discovery of that one simple drug, antibiotics are an entire class of medicines with a global impact so profound that we take it for granted. Invisible terrors that plagued and poisoned mankind for millennia are now casually vanquished at a drive-through pharmacy.

The discoveries of how cancer tricks and hides from the immune system were immunotherapy's penicillin moment. The approval of the first checkpoint-inhibiting drug that regularly and profoundly changed outcomes for cancer patients redefined the direction of scientific inquiry. That's now kicked off a gold rush in research and investment and drug development. Seven years after the approval of that first solitary checkpoint inhibitor there are reportedly 940 "new" cancer immunotherapeutic drugs being tested in the clinic by more than a half million cancer patients in 3,042 clinical trials, with another 1,064 new drugs in the labs in preclinical phase. Those numbers are dwarfed by the number of trials testing the synergetic effectiveness of immunotherapy combinations. The research is advancing so rapidly that several drug manufacturers have successive generations of drugs stacked up in the clinical trial pipeline like planes waiting for clearance at LaGuardia, requiring new FDA "fast track" and "breakthrough" designations to speed them through the approval process to cancer patients who don't have time to wait. Major advances in cancer usually come in roughly fifty-year increments; cancer immunotherapy has already made a generational leap, seemingly overnight. Describing what is coming next, many scientists smile and use words like "tsunami" and "tidal wave." The pace of progress is rare in the history of modern medicine, unprecedented in our history with cancer. We have an opportunity to fundamentally redefine our relationship with a disease that for too long has defined us.

This is the story of the geniuses, skeptics, and true believers, and most especially the patients who spent their lives, and the many more who lost them, helping refine and verify this hopeful new science. It's a journey through where we are, how we got here, and a glimpse of the road ahead, told through some of those who experienced it firsthand, and some who made it possible.

Chapter One

Patient 101006 JDS

Scientific theories...begin as imaginative construc-
tions. They begin, if you like, as stories, and the purpose
of the critical or rectifying episode in scientific reasoning
is precisely to find out whether or not these stories are
stories about real life.

—Peter Medawar, *Pluto's Republic*

Jeff Schwartz's story begins in 2011, when researchers had discovered some of the secret handshakes cancer uses to trick our defender immune cells. Newly invented treatments blocked that handshake and unleashed the defenses in our blood. These drugs were available in trials, but not everybody knew about them.

Many cancer doctors were unaware of the new developments that might save their patients' lives. Others refused to accept that such a breakthrough was possible. That refusal denied their patients the option to try it. It sometimes still does. It's why Jeff Schwartz was willing to share his story.

~

Jeff Schwartz knows he was one of the lucky ones. His father died from lung cancer in the '90s after increasingly ugly attempts to beat

9

it—the usual protocols of cut, poison, and burn; surgery, chemo, and radiation. Just before the spring of 2011, Jeff was diagnosed with cancer too, kidney, stage 4.

So Jeff considers himself lucky, or blessed, or—he doesn't really like to put too fine a point on it, you know? It wasn't because he had some sort of influence or special knowledge or anything of the sort. What separated Jeff from the hundreds of thousands of people who died of the same disease during the same time was that he happened to live in California and happened to walk through the right door at the right time. That's changed the way Jeff thinks about life and living. Now he hopes his story might reach someone else so they don't have to be lucky.

I met with Jeff in his room on the forty-third floor of a hotel in midtown Manhattan. Jeff looks a little like a more biker version of Billy Joel post-booze and post–Christie Brinkley. He was dressed in jeans and a blue Izod shirt that hid the hard edges of the titanium cage that prevents his spine from collapsing. Surgeons had implanted it there, like he was Wolverine, after tumors had eaten away his spinal architecture. He told me about the cage. He pointed out the scars. These were facts, that was all, part of the story he was telling.

Jeff Schwartz had been a kid in Rockaway, Queens, who went through the public school system and drove a cab while he earned a degree in accounting and economics. His first job was at the mortgage desk of Lehman Brothers, his next at a Japanese bank run by Harvard MBAs. Neither was a good fit. Jeff was a music guy. He played guitar "pretty good," he says; it was his secret identity, the *other* thing you tell people at parties when they ask what you do: "I'm an accountant, but really _____." And maybe for good measure Jeff might talk about any of the hundred-plus Grateful Dead shows he'd seen, or how he'd been given Allman Brothers tickets for his bar mitzvah, or show you the first two measures of John Coltrane's "A Love Supreme" tattooed around his left ankle like a musical mandala. Nights after the trading floor closed he'd head to the East Village to mix boards at CBGB and the Mudd Club for Talking Heads, Blondie, and Richard Hell and the Voidoids—he's especially proud of assisting in the recording of *Blank Generation*. Maybe he wasn't cool, he says, but he was on the scene.

His passion transitioned to his career because of baseball. He'd done a favor for a guy, and the guy thanked him with a pair of expensive tickets. Jeff had been a diehard Yankees fan his whole life. These were Mets tickets, great seats, really wrong team. So Jeff gave the tickets to a friend, who invited another friend and, long story short, that friend made him a job offer to be a junior guy with his company, a financial firm for clients in the music business. Jeff would come on as the young guy to help with the young talent. His first client was a new girl act, Joan Jett. That worked out for a few years, exciting times, and eventually he opened his own shop and moved out to Malibu.[1] His wife was a record company exec, they had a kid, they had a Lexus. He had an eye for talent and made 5 percent of what he made his clients,[2] and when one of his acts blew up, like Ke$ha or the Lumineers or Imagine Dragons, Jeff made good. But the real perk was his access. Stopping by their live shows was the coolest counterbalance to careful spreadsheets and checked math.

He admired the musicians, dug the music. But his value lay in the practical side. Music is a profession, though many musicians fail to realize it until it's too late. "Most acts are one-hit wonders, guys smoking pot in their dorm room, they come up with a song that happens to be pretty good and then, that's it," Jeff says. "I tell

my acts: If you don't want to be serious then you're wasting everyone's time. Yeah, be a rock star, but this is how you're going to buy
your house. It's going to be your retirement account. It's how you'll
probably meet your wife or your husband. It's more than lifestyle,
it's your life." As far as he was concerned, the song you *wish* you
wrote? It isn't "Yesterday." It's "Tie a Yellow Ribbon 'round the Ole
Oak Tree." They're both about remembering, but only one made a
billion in Muzak covers alone.[3]

Jeff helped with the contracts, advised on royalty deals. There
were writing fees and pennies from records or plays on streaming
media, iTunes, Pandora, Spotify—the music world was changing fast in the early 2000s, and you had to watch every stream.
The more digital the music got the more free it got, and the more
it served as advertisement for the payoff of an international tour.
Sending an act off was like christening a new trade ship after years
of building. It could make or break, and Jeff wanted to be there.

And so in February 2011 he was in Portland, Oregon, watching the roadies set up for the first night of Ke\$ha's new tour and
wondering if maybe he was pushing himself too hard. The 2011
"Get \$leazy" tour—the dollar sign in place of the *S* being Ke\$ha's
trademark—had packed shows scheduled across the Americas,
Europe, Australia, and Japan. Jeff had taken Ke\$ha on when she
was a kid playing club dates. She'd blown up when Rihanna had
signed her to open on her world tour, and now, at twenty-three
years old, she was positioned to leave port and capitalize on the
zeitgeist, with Jeff on deck to help steer the finances.

Jeff didn't need to show up, but his presence there was a personal reminder to his talent. He was looking after their investment,
and that investment was themselves. They should do the same. He

really couldn't avoid opening night, no matter how he felt. Which was too bad, because Jeff was feeling like crap.

He was always a little sick these days, a little weak, more than the usual morning stiffness—and the general ache now lasted all day. That came with hitting fifty, he knew that, the way his hair had gone white and thinned at the top. He'd adapted, wore it cut short with a white goatee. Late nights and discomfort were part of the rock-and-roll soul swap, same as the inevitable weight gain of late drive-through meals and no exercise. At least there was an upside—between the pain and the nausea, he was losing weight. He hurt, but he looked good. When he hit 180 pounds, he was happy to recognize his old silhouette in the hotel mirrors. But the weight kept dropping and he felt something else, a dread he couldn't put his finger on.

Ke$ha, decked out in a rhinestone-studded leotard and laser-shooting sunglasses, ducked into the spotlight. Jeff felt cold. There was a pain in his side, or his belly, or his back—somewhere in the middle there. He wasn't feeling any better as Ke$ha came back out in a star-spangled getup and fishnets to sing her hit, "Fuck Him He's a DJ." Jeff found a seat and watched the backup dancers and the band, professional musicians whose costumes were described as a cross between "Mad Max and prehistoric birds." It was nearly midnight when Ke$ha finally performed a lap dance on an audience member duct-taped to a chair. An extra in a giant penis costume pogoed around the couple in a choreographed number.

Jeff checked his watch. The encore was thundering. Thank you, Portland, Oregon, and good night. Maybe, Jeff thought, he needed to just go lie down. But the pain he'd felt had been at a different level, and it didn't go away. Ke$ha's buses headed off to the next stop on the tour. Jeff stayed behind and quietly drove himself to the hospital.

A doctor looked him over. A phlebotomist took his blood. They ran the numbers, brought him back in, asked him to sit down. He remembers the doctor telling him that the first thing that stood out was his hemoglobin count. It was staggeringly low. With numbers like that, his blood didn't have the means to transport oxygen to his muscles or his brain. That was probably what accounted for his exhaustion. But what accounted for the low hemoglobin? It might be cancer.

That suspicion led Jeff to the Angeles Clinic on LA's Wilshire Boulevard—PET scans, the usual round of tests—and on President's Day weekend Jeff was told: kidney cancer, stage 4. He didn't know about stages, but he did know there wasn't a stage 5.

He also didn't know, and in the moment of shock wouldn't have cared, that he was one of about sixty-three thousand people in the United States to get a diagnosis of kidney cancer that year. Of those, a far smaller percentage would get a diagnosis for the rare and specific cancer Jeff had. It was, in the language of cancer specialists, an especially "interesting" type of cancer, a particularly aggressive variety called sarcomatoid renal cell carcinoma.

"The doctors tell you, don't go online when you get your diagnosis," Jeff says. There's no good that can come from trusting everything put up on the internet to interpret your fate. "But of course, that's exactly what you do."

He got as far as his car. He took out his phone and looked. The numbers, at first, looked—not bad, really. The five-year survival numbers, the standard numbers given for cancer at that point, were nearly 74 percent. That's a passing grade, a majority, Jeff remembers thinking.

But then, reading further, he saw that the good number depended on other factors. The most important was how early you caught the disease.

The kidneys sit in the lower back, two filtering masses about the size of fists on either side of the spine, right about where you might hold someone to slow-dance at a junior high prom. They're complex filters, composed of millions of tiny, capsule-shaped glomerular filters that sort out what the body needs from what it must discard. But like a demolition worker clearing out asbestos from a condemned building, those glomeruli are heavily exposed to all the concentrated toxins that come through the body. They are more likely to undergo DNA mutation as a result of that exposure, just as exposed skin catches more UV radiation and is more subject to the mutations that facilitate melanoma.

The survival rates Jeff was looking at were when you caught it early—when it was just in the kidney and the tumor was no larger than seven centimeters.

The United States doesn't like metric measures, so it tends to translate them to nuts and fruits, and sometimes eggs and vegetables, to describe tumor size. For a five-centimeter, stage 1 tumor, the American Cancer Society site uses a lime. Stage 2 is a lemon, or a small orange, still localized as a mass within the kidney. Stage 3 means the tumor has started to spread within the kidney. The growing, spreading cancer—a peanut, a walnut, or an orange—if it's stage 3, is still contained within the kidney area, so it can be more readily targeted by conventional cancer therapies—specifically surgery and radiation.

Since most of us have two kidneys, and can survive on one healthy functioning one, cutting out one whole kidney—what they call a radical surgery—is a common approach. But Jeff's diagnosis was stage 4. That meant the tumors had entered the bloodstream and moved elsewhere, and possibly everywhere.

No matter where those mutated renal cells moved—they could fill the lung, lodge in and take over the liver—they would always be called "renal cancer." (This naming system, as anachronistic as describing tumors in terms of fruits, changed because of cancer immunotherapy in 2017, itself a breakthrough.) And so when those mutated renal cells started colonizing his spine, Jeff's cancer was "kidney cancer," stage 4. And on the tiny screen of his flip phone, stage 4 kidney cancer looked really bad. The five-year survival rate hung at a guttering 5.2 percent, and it had been about 5.2 percent since the 1970s. The last new scientific advance for treating kidney cancer had been made thirty years ago. There isn't any way to put a positive spin on that. You just close your phone, sit in your car, and wait until you're calm enough to drive.

There really isn't a good time to get a diagnosis like that, Jeff knew. Jeff was busy—but everyone is too busy for this sort of thing, and once he went through the usual reactions he realized that too. But hey, come on. He was *really* busy. His business was booming, his acts needed him, and he now had two little kids—one three years old, the other just a year. He wasn't going to stop working, he wouldn't make a big thing of it. He told only those clients who really needed to know, who'd need to make professional decisions. He told Kesha he was sick, didn't say how sick. That seemed OK. Mostly, he decided to just keep moving forward.

Next, Jeff was referred to the larger affiliate hospital, the mothership, to see their kidney specialist. Maybe it was Jeff's mood, but this doctor, he decided, was "a fucking prick."

Let's call him Dr. K. He had looked at the numbers. Stage 4

kidney cancer was pretty much a death sentence, especially in this rare aggressive form, but there was always a chance. Dr. K started Jeff on a drug called Sutent. As the label promised, Sutent gave Jeff the usual symptoms of extreme nausea, lack of appetite, and daily dry heaves.

Meanwhile, his PET scans had come back. The cancer in his right kidney was now working its way up his spinal column, tumors leapfrogging each other like kids grabbing a bat handle for dibs. They scheduled surgery to look at that, and when the surgeons opened him up they found that the tumors had eaten into the bone. Fists of dense tissue fissured the central supporting column of his body and nervous system, and knuckled perilously into the wiring of his spinal cord. The structure was brittle and laced with a progressive disease; soon the tumors would either engulf and seize his spinal nerves, or his increasingly fragile vertebrae would fail under his weight like the collapsing World Trade Center towers, or maybe both.

It was moving fast, and either scenario would leave Jeff a quadriplegic at best. They needed to immediately secure the structure. The cancer was inscrutable, incurable, and complicated, but this was concrete physical work that a surgeon could do with a knife. Chunks of his spine needed to be cut out and titanium rods screwed in their place. It would give Jeff a Frankenstein posture, and he would have to live with a constant hum of background pain from the raw nerves—compressed by his collapsing vertebrae, permanently pinned to the titanium infrastructure like guitar strings hard against a fret board—but at least it kept him from being paralyzed. It was the deal to make. A month later they operated again, and finally took out the diseased kidney.

It was hard, the surgeries and the pain were extreme, but "I never stopped working," Jeff said. "I tried to hide it from everyone." He still got up in the morning, showered and shaved and dressed, cinched his belt tight to keep his pants from falling off his hip bones, got in his Lexus, and headed toward the freeway as he'd always done. Going to work.

"But I never went to the office." Instead, he'd pull off somewhere south of Malibu, go through the McDonald's drive-through to get an Egg McMuffin, and force it down before he even pulled back out to the road. Then he'd drive up and down the Pacific Highway taking calls on his car phone. "Every once in a while, I'd pull over, put the phone on mute, throw up out the window, and get back to the call," he said. The McMuffins helped; they were soft, and much better than the dry heaves.

He had two doctors: Dr. K, his kidney specialist; and the surgeon, Dr. Z. Jeff saw Dr. K for the sutent, and met with his surgeon a few weeks later for the follow-up. Both doctors had seen the same scans but gave him different messages. "The surgeon told me not to bother with the chemo anymore," Jeff says. "He thought I should just give up trying to beat this thing and try and enjoy the little time I had left, without the side effects." Dr. K was upset that the surgeon was telling his patient to ignore his prescribed treatment.

As far as Jeff was concerned, it wasn't that Dr. K disagreed with the surgeon's prognosis; he, too, thought Jeff was going to die. It was that the doctor was getting paid for every chemo treatment Jeff underwent, and Dr. K wanted to keep collecting the fee for the treatments, as long as Jeff was alive enough to take them.

Finally, in September, Dr. K gave him a final prognosis. "He told me that I was going to have six months, tops," Jeff said. In retrospect, it's surprising he was given that long. Jeff's weight had dropped to 148 pounds, and increasingly more and more of that was tumor.

"The guy told me to go get my financial affairs in order," Jeff says. "He was awful—no bedside manner, no compassion." The way Jeff read it, "They were done with me. They'd given up."

Jeff believes it was a matter of the doctors at the hospital having nothing left to bill for, that was how he saw it. Maybe that's just how he thinks as a manager and accountant; maybe it's more than that. Doctors are just people. While the best of them are very good at aspects of their jobs, and some of them are good at several aspects, it's rare to find one who can act as both an expert physician for a patient's physical body and a philosopher-priest for a human mind contemplating its own death. These are also the desperate thoughts of a desperately ill man raging against the tyranny of mortality, as presented by a series of guys in white coats. It's a tough go, however you look at it. Bad news isn't easy on anyone.

Either way, the medical professionals, the ones who knew more about what was laying waste to his body than Jeff himself could comprehend, had nothing more to offer. They saw no option but to give up. And so, the logical thing for Jeff to do was to follow the experts' lead and give up too.[4]

His referring physician at the Angeles Clinic, Dr. Peter Boasberg, had another idea. There was a clinical study. Maybe, possibly, he could get Jeff on this study. "Maybe" sounded pretty good at that point. The drug being tested didn't attack the tumors; rather, it attacked the tumors' ability to shut down the natural immune response against them. It was called a checkpoint inhibitor. Already, there was a theory among researchers working with this drug that it was most effective in generating a strong immune response against tumors that had a high degree of mutation. That might include kidney cancer—cancer like Jeff's.

~

All decisions about the study parameters fell to Dr. Dan Chen, MD, PhD, an oncologist immunologist who was also the cancer immunotherapy development team leader at Genentech, the company that made the experimental drug. Patients that might qualify for the study were presented to him in blind applications—each reduced to a code name of letters and numbers and the specifics of their medical history. Jeff Schwartz was now patient applicant 101006 JDS.

Originally the drug study had been designed to look at the drug's effectiveness against solid tumors. It had been expanded to include melanoma, bladder, kidney, and several others. Did Jeff qualify for such a study? On paper, the answer wasn't obvious.

If Chen was looking for reasons to exclude someone, he could definitely find justification to rule out patient 101006 JDS, but that didn't make it the right decision. The study qualifications were advertised in terms of what types of cancers qualified. That description did not specifically mention the rare form of kidney cancer on patient 101006 JDS's paperwork—but it *was* kidney cancer, and Chen strongly suspected that 101006 JDS's rare cancer had many similarities to the cancers they believed would be responsive to their immunotherapy candidate, a new checkpoint inhibitor. On the negative side, this rare and aggressive cancer was now rooted deep in the bone, which the immune system has a difficult time infiltrating, but this patient fit the profile and, Chen suspected, might benefit from the experimental drug. If it was already approved and readily available, Dr. Chen would have put him right on it and hope it helped, as nothing else had. But in 2011 this immunotherapy was not yet an option in an oncologist's toolbox. The only way for a cancer patient to access this drug was through the experimental trial.

Which made patient 101006 JDS an especially tough call. Chen knew the usual course of a stage 4 kidney cancer prognosis. As a physician and a compassionate human being, he wanted to say, look, if 101006 JDS's disease qualified, he was in. But as a scientist and department leader in charge of a massive phase study, there was a problem. Based on the paperwork, patient 101006 JDS was probably too unhealthy for a physically rigorous trial; he might jeopardize the whole study. There was no computer algorithm, no chart or slide rule for making this decision. Chen had to balance out the factors and weigh them with his head and his gut.

~

Jeff didn't know how to play the odds on this one, how hard he should hope or how he should proceed with the next phase of his life. On the one hand, there was no guarantee he would be able to start the new experimental drug trial for the immunotherapy, and he should prepare for that. But on the other, if he *did* beat the odds and get a green light, he'd need to be available to accept on the spot. To do that, he couldn't be on any other cancer treatment. That meant he'd have to stop the chemotherapy and wait. The chemo had been making him feel terrible, and it hadn't stopped his tumors from growing, but it was the only treatment offered. There was no telling how quickly his cancer would progress if he wasn't actively poisoning it with the chemicals. It was possible that the chemo was slowing down his decline and buying him a few extra precious days or weeks with his family. Quitting that possibility on the slim chance that he *might* start something else, something experimental, that *might* work, and might not—that was a perilous trade-off. It felt like holding your breath to avoid breathing in poison.

~

Years later, Dan Chen still remembers everything about patient 101006 JDS—his profile as a potential study participant, his response, even his coded patient identity number, straight off the top of his head. As a scientist testing his first immunotherapy drug, Chen wouldn't quickly forget his first responder. 101006 JDS turned out to be, as Chen says, "a special, special case." Part of what made him special was the fact that, even knowing how it turned out, one could *still* make the case that from a data collection perspective, patient 101006 JDS shouldn't have been allowed onto any kind of medical study at all.

"My initial reaction upon seeing him—it was all just on paper at that point, but it was, 'Are you kidding me? Why are you sending me patients like this?'" The study Dan was running was a phase 1, meaning the first in humans, and his team had something of a fire under their ass to make this thing go. They'd started late in the immunotherapy game, despite the number of cancer immunologists who were then like secret sleeper cells within Genentech, and when they'd convinced the larger company to change the course of their research and allow them to head into this unproven direction for their drug development they were already years back, building a new drug program from scratch.

Before he'd joined Genentech, Chen had been working with cancer immunotherapy both in his lab at Stanford and with his patients at the Stanford University Cancer Center. Those early approaches hadn't worked against cancer. But despite the failures of the various vaccines they had tried, and the uneven and sometimes disturbing effects of giving patients doses of strong immune stimulants like interleukin-2 and interferon, the researchers had seen glimmers of hope. Chen and other cancer immunology believers had seen them in their rare but real positive responders, and in those reported by a handful of other labs around the world. Most oncologists—most scientists—dismissed cancer immunology as a dead end, peopled by quacks and true believers who confused hope for good science. But Chen believed, as the handful of others who were still in the immunotherapy field believed, that there was something more to these positive responses than misinterpreted anecdotes.* This drug study might help prove that.

Was patient 101006 JDS going to help the study? Chen wasn't so sure. "He had a lot of disease. The disease was in bad places, including the bones, which is harder for cancer immunotherapies to treat," Chen remembers. Worse, his "performance status" was horrible.

Performance status means: "What's your day look like? Are you up and about? Or are you unable to get out of bed because you're puking all day, and can't eat?" Chen and other oncologists used performance status to predict how a patient was going to

---

* See Appendix B for more information.

fare—whether in a clinical trial or with a traditional cancer ther-apy. It's a more rigorous variation of: How you doing? And 101006 JDS wasn't doing well.

"If you can't get out of bed, if you can't move, your outcomes are generally horrible," Chen says. "Sometimes you have patients that are declining like this—" He puts his hand out, tilted like a graph nosediving. "You generally can't reverse those patients. So putting people who are on a downward course on your trial is not a great way to figure out if your drug is safe."

And that was the point of this phase 1 trial—to gauge the safety of a potential new drug by testing it in low doses. If it failed here, it failed. If that test was going to mean anything, it needed to be the truest possible reflection of the safety of the drug. From this per-spective, patient 101006 JDS wasn't exactly who you were dream-ing of. Patients who were too weak and sick would fail the test no matter what you gave them—and that failure would be ascribed to the drug, not to the patient. It wasn't just Jeff who would suffer, it was the entire study and, by extension, an entire generation of patients.

If the "How you doing?" test was somewhat subjective, the main criteria of entering the study were standardized and empiri-cal. "We had a laboratory value that you had to meet," Chen explains. These values had been provided to all the primary inves-tigators who would be running the study; their patients needed to meet or exceed those values in order to even be considered.

101006 JDS's lab values were *bad*. His albumen, his white blood cell count, was "not good." Those values were a particularly nega-tive indicator for a potential immunotherapy drug study candidate.

"First of all, you need to have the white blood cells," Chen says. "You need to have the T cells. We didn't know much about the drug at that point in time, but if you don't have the T cells in the first place, then why would we be trying to give you a drug that's going to react with the T cells?" That was the most important number in the trial, and Jeff was below that number value. "There was no way."

Two months off the chemo, Jeff was sicker than ever—too sick to meet the eligibility requirements for the Angeles Clinic study.[5] Thus began a back-and-forth, with Jeff's doctors—Chen's primary investigators on the study—in the middle.

"They had protocols," Jeff says. "My hemoglobin had to be at a certain level. They drew my blood; I said, 'Draw it again.'" Maybe Jeff's levels were fluctuating, "so they'd try drawing it at different times of the day," he says. "I was eating broccoli like crazy, every day, just trying to get the numbers up."

"I know they were really trying anything they could," Chen says. "There had also been an old observation that showed that rubbing the earlobes pushed out white cells—a real phenomenon, studied at Johns Hopkins, called something like 'earlobe lymphocytosis.'" So they tried rubbing his earlobes. Jeff rubbed them at night or in the car, rubbed them right before they drew blood. That didn't get his number up high enough, either.

In November Jeff's oncologist at Angeles, Dr. Boasberg, had to break the news. He wasn't going to qualify for the study. "And I knew that was a death sentence," Jeff said. He wasn't ready to give up, but he couldn't simply will his immune system into health. "They offered to put me on a different drug trial," he said. It wasn't an immunotherapy, and Jeff had already gone the chemo route. It hadn't worked, and it made him feel like shit, and he maybe only had a few months anyway. Was he really willing to feel like that again?

Absolutely, if there was a chance it would help him, that was his attitude. He hadn't tried this particular drug, so at least he could try to think of this as plan B. But maybe, Jeff worried, it wasn't really a plan at all—it was simply something to do, the medical equivalent of busywork for the doomed.

You try to put a good face on it, to go along, be a good patient, to not look at the couldas and wouldas. Cancer is full of those— lung cancer clinics filled with smokers who quit—and the important thing to Jeff was to keep moving forward. But it was tough not to see that two roads were diverging here, not to recognize that plan B was the wrong fork. The thing that his doctor wanted for him, the thing he thought might possibly help him—Jeff couldn't get. This other study was clearly an afterthought—but maybe busywork was what he needed. Maybe it would help him. Definitely Jeff wasn't seeing any other choice that didn't amount to giving up and making peace with his fate.

The only problem was, Jeff wasn't feeling peaceful. He kept glancing over his shoulder at the other path. Now, it was Jeff who had the tough decision: starting the study on offer meant giving up on some miraculous possibility of entering the checkpoint inhibitor study, but this bus was leaving, too. If he waited, he'd be left at the crossroads, stuck with nothing. And nothing, he'd already been told, meant entering hospice.

~

Meanwhile at Genentech's San Francisco campus, Dan Chen had a problem. A number of them, really. One was a problem every oncologist had, the burden of the job. Cancer treatment is generally not a good-news field. To be a good doctor and researcher, you had to accept the fact of mortality and the terrible outcomes from the disease, even while working actively, often fruitlessly, against them, every day.

Part of what he had to accept regarded the fate of potential patient 101006 JDS. He looked bad on paper, but he'd been wavering near the line long enough that, even behind the coded identity, his case had become personal. Dan hoped for a good outcome for this guy—he knew it was a man at this point—but he also hoped for good outcomes for his drug, and for cancer patients in general.

"And now we're coming up on the Christmas holiday, and everything is going to shut down," Chen says. There would be a break at the company, short breaks maybe for some physicians, and the patients themselves might choose to go see far-flung friends and family, some of them for the last time. It's what happens. That meant that the race to get this drug developed, tested, and to patients and the marketplace was about to suffer a major delay. "So I had to face the fact that if we don't fill out this cohort before the holiday, we delay this whole trial," Chen says. And that would have a ripple effect, and potentially serious consequences. The other patients couldn't start until the cohort was filled. And no patients could get the drug, or have reason to believe it was worth getting, until the drug passed through clinical trials and, best case, expedited FDA approval. That one empty spot in his cohort had become a stop sign.

If patient 101006 JDS was ever going to be considered, it was now or never.

∾

Jeff Schwartz was scheduled to begin the plan B clinical drug study on December 17. He remembers the morning, the car, the highway.

The drive to the Wilshire Boulevard clinic was like a gallows march, dead man driving. The windows of his Lexus were shut tight against the Los Angeles air; his heater was cranked up to eighty degrees, just to keep him from shivering too hard to drive.

"I didn't tell anybody, but—that was awful," Jeff says. "I'd sorta resigned myself to it. I was going to keep fighting but…" Jeff stops. He hadn't thought further that day either, at least not about himself, because, as far as he was concerned, he was done. The rest was about fiduciary responsibility. "Every penny I made, I made sure my kids had savings," he says. "I paid my leases up front, not knowing if I'd be there to make the next payment. I didn't get spiritual from it, it didn't change that, but death, the prospect of it, knowing the end—" Jeff stops and considers the room for a moment. "Well, it changes the way you think about things."

Jeff drove through the gate. He parked, pulled himself out of the car, checked in at the desk. There was the clipboard with the cheap pen. A nurse came out, called his name. She waited, smiled, turned. He followed through the doors and into a room, a comfortable chair. The overhead lighting was bright. His name had been transcribed to an ID, then matched with an IV. Studies have to be double-blind to avoid any bias or sentiment. For the sake of science, the patient is stripped of his identity. Good for science, but hard on us humans. The procedure would be: pull out the drip, check the numbers on the drip against the bracelet on the patient, enter the numbers on the form, hang the IV, unstop the stopcock. Patient 101006 JDS, formerly known as Jeff, was ready. His sleeve was rolled up, a catheter needle was inserted and taped. The drug would be useful, they'd discover later,[6] for giving extra months of life to some kidney cancer patients, but almost certainly not of any use for patients like Jeff.

~

Five hundred miles north in San Francisco, Dan Chen was in his office as the sun came up. At 7:30 his phone rang. The sudden noise startled him. The call regarded patient 101006 JDS's latest numbers. Maybe it was the ear rubbing, Dan doesn't know; maybe there's a value for sheer will. Whatever it was, the guy had popped over the line.

Maybe the numbers wouldn't stay there, but they'd run the tests and he'd passed. It was a cold line and he'd crossed it. That part wasn't a judgment call now, it was empirical. The next call, if Dan made it, would be to the clinic: Can they put this guy on the trial?

Dan remembers the light. It bounced off the cold gray chop of the San Francisco Bay and it did something to the room. He watched it, looked out the window.

"It wasn't just about this one guy," Chen says. "The trial would affect far more people. Was this guy just going to crash and burn?" Would he hurt the trial and hurt those people? If he let this one in, was Chen doing the right thing, or the wrong thing?

It all had to happen in a matter of minutes, the call with the numbers, his decision, but something about the light—maybe he'd watched too many movies but there was something of a Christmas miracle in it, that feeling you get that time of year, maybe a kindness, even when kindness could be misguided. Dan picked up the phone, called the clinic. The line was busy. He put the phone down, checked the number, tried again. Same thing. Maybe this was a sign? Maybe it was just a busy phone. He tried once more and got through. He gave the patient number, said, "Let's put him on." There was a pause and a noise, and a sort of panic. It sounded like somebody was running.

~

"So I'm sitting there," Jeff said. "They had everything hooked up, that was it. The bag was hanging, they just needed to finish the job. And a nurse came running in and said, 'Wait.' Like there was something wrong with my blood or something."

Then the doctor came in. "They called," he said. Apparently, Jeff's lymphocytes were up enough to fit into the protocol for the study.

"They said they were 1100 or something," Jeff says. "I had the same cancer as before, it wasn't getting better, but my numbers were better." According to the blood work, his lymphocytes had finally showed up. "Maybe there was some miracle, what happened—I don't know." What he does know is they pulled the drip, unhooked the line from his vein. His doctor had one final message from Dr. Chen. "He said the message was, 'Tell the patient Merry Christmas.'"

~

Three days later, on December 20, Jeff became patient twelve of the twelve-person study. He drove himself to the center. It was the first time the clinicians administering the study had met their trial patient in person. He was sicker than they had imagined. Sick enough that they called to make sure this guy was really supposed to be there.

Jeff went through the same setup as before: the clipboard and the forms, the stickers and the armband and the rolled-up sleeve and the needle. Except this time, he was infused with the experimental checkpoint inhibitor immunotherapy drug.

This was Jeff's first experimental drug. They called it MPDL3280A. The mad scientist notion of that, *an experimental drug*, that was exciting, but also a bit frightening. MPDL3280A had worked in mouse models, but 90 percent of all cancer drugs that work in mice fail in human trials. "I asked them, 'Hey, this stuff you're giving me—is this going to blow my head off?' And they say, 'Fuck if we know.' Because I'm the first one!"

It didn't blow his head off. But it did do something. "Right away, I just came back to life," Jeff says. "It was weird." Was it really working? Or was that just in his head? He'd only gotten one dose, and a low dose—that was one of the goals of a phase 1 trial, to find the "lowest effective dose" of a new drug. That seemed like an unlikely scenario for an instant effect.

Jeff knew about the placebo effect, and he knew about the effect faith and hope can have on a person's health and even on their outcome. He did plenty of that with his clients, talking them up; the power of belief was important and real, but it didn't cure cancer. He also realized that he'd been off the chemo long enough that, no matter what, he was going to be less nauseous.

The next visit, two weeks later, they injected him again. And again, right away, he was pretty sure he was feeling better. He'd always been working, even while he was dying. Now he felt good enough to do things besides work. Good enough to take his five-year-old son to SeaWorld that month.

"I'm there, I felt a pop in my hip—it was my hip bone. The cancer had eaten right through it, it just popped right through the hip socket." That did nothing to help his performance status, so that was a setback. He had another surgery, but it was just a surgery, rather than new cancer, one of his old problems coming home to roost. You didn't worry about the old problems; you hoped to stop any new ones. His next appointment comes, Jeff gets another infusion of the experimental drug. And this time, he doesn't just think, he's *sure* he's better now.

At home a few weeks later, his son asked him: "Dad, what

happened?" Jeff didn't know what he meant, but his son told him—he didn't think that his father could lift him anymore. But here he was, throwing the boy up in the air, watching him squeal with delight. Jeff hadn't really thought about that, but his son noticed. *Something* was improving. Then, the PET scans confirmed it.

On March 15, 2012, Dan got an email update from the physician at the trial clinic. The iffy patient they'd sent over, the one with significant fatigue, pain from a retroperitoneal node, the one unable to work or lift his young children—they wanted Chen to know. Patient 101006 JDS had been "recalled to life."

~

Jeff understood how lucky he was, and he wanted to meet the people responsible for what he had received. He had known that the global lead on the study was a Dr. Chen, based somewhere in the biotech boxes in San Francisco. He'd hoped to call in, maybe a conference with the team, just to say thank you to the room.

"In July they told me there was a conference in LA, and that 'Dr. Chen' was going to be there. I'd always pictured this Dr. Chen, the mysterious Dr. Chen—you know, I'm picturing some geek in wire-rimmed glasses somewhere. I meet him and, here's Mr. GQ."

Jeff found Chen surprisingly charming and accommodating. Later, Chen brought Jeff into the San Francisco offices, then down to the labs. There was a sort of Willy Wonka feel to the whole thing. "This place, they've got a sign reading E. COLI on one side, CHO on the other," Jeff says. "I ask him, 'What's "CHO"?' He says it's 'Chinese Hamster Ovaries'! Then they bring me over there to a big steel vat. He asks me, 'Know what this is?' I tell him, 'It looks like a brewery.' Dan says, 'Yeah, well, they're brewing proteins.'"

Finally, Jeff was introduced to four researchers who had helped develop the drug and build the protein.[7] "I meet them, they know who I am, and so they're all crying," Jeff says. "Because, these guys are geniuses, but they're all like idiot savants, they never leave the lab." After all the work that had gone into getting their immuno- therapy drug ready, and after decades during which immunother- apy drugs had failed to save patients, the sight of a healthy man, returned from the brink of death by virtue of their efforts—it was a first.

"All these guys, I don't know how they do it. They get rejected all day, everything they do fails, people die. Can you imagine being these guys? Chen is an oncologist, he does melanoma, all his patients, all these guys. Can you imagine?"

Dan Chen had done something for him, something big. So Jeff wanted to do something for Dan. "Chen doesn't know me, not per- sonally, he knows me by my initials from the study. So, he doesn't know I'm in music." Jeff asks Dan, who's got a teenage daughter,[8] "So, what's her favorite band?" Chen tells him it's a band called Imagine Dragons. Jeff smiles as he recalls: "So, next time the band is in San Francisco, I get her tickets, she goes backstage, then she's on the stage throwing out balloons—it's great! She's having the time of her life. Dan says, 'Thanks.' I tell him, 'Hey, thanks for keeping me alive!' "

∽

Jeff was feeling well enough now that it was almost life as usual. That included going to his son's weekend basketball games in the grade school gym. "My wife's with me, she says, 'Look, you know who that is, across the gym?' I look and I can't believe it—it's K. Dr. Fucking K.

"So I march right over, I ask him, 'You remember me?' He says no. I tell him, 'Well, I'm the fucking prick you gave five months to live!'

"I saw him again six months later—our kids are the same age at the same school, he's going to see me, and I went up to him *again*. I told him—I just had to say something. I had to tell him, 'Look. A patient hangs on every word. *Every word*. You told me you had nothing left, told me I had five months to live, and my world was shattered.'"

Jeff also remembered what his other doctor had told him when he was getting sicker and hadn't found a clinical trial. Dr. Boasberg had seen the new immunotherapy drug trials come through, he'd seen the transformative effects, things they never imagined. And he knew how quickly the new drugs were coming. "I might not get into the study now," Jeff said. "But he told me, 'Hang in there, because there are new drugs right around the corner.'"

In 2011, that was a rare and radical perspective. Not every oncologist was aware of what was happening in cancer immunotherapy. At the time, the vast majority still thought of the immune-based approach to treating cancer from the bad old days of false promises and ineffective vaccines. Jeff was lucky enough to have a physician connected to a place like the Angeles Clinic, staffed with oncologists open to the potential of cancer immunotherapy and geared toward clinical trials. Of course, he'd only gotten there because his cancer failed to respond to anything else. In that sense, he wasn't so lucky.

When Jeff was in his twenties, cancer was an old man's disease, and he didn't much think of it. He was the kid filing out from some live show downtown with his ears ringing as snowflakes slanted through the streetlights. Not to get sentimental, but that trouble-free time really did seem like yesterday, like in the song. He can still picture the snow landing like lace on his black leather jacket, there and gone. He had all the time in the world.

"Just think of all the people who died, waiting," Jeff said. "Just a little too early, or the ones who just gave up because a doctor told them they were done, that's it." That could have been him. It wasn't. And why was that?

Jeff doesn't know. But part of it has to do with luck, part with sheer will, part with faith, or something like it. And part of the answer took place more than one hundred years before, on the same downtown streets of Jeff's youth, where a New York surgeon chased a medical mystery into the immigrant slums and returned with a recipe to cure cancer.

Chapter Two

A Simple Idea

The real voyage of discovery consists in seeking new landscapes, but in having new eyes.

—MARCEL PROUST, 1923

In terms of modern Western medicine, the idea of using the body's own immune system to kill cancer traces back through the end of the nineteenth century and a seventeen-year-old girl named Elizabeth Dashiell. "Bessie" was the pretty and self-possessed daughter of a Midwestern minister's widow. She was also the very close pal of the only son and namesake of the founder of Standard Oil, John D. Rockefeller Jr. There's never any mention of romance in their relationship—Rockefeller referred to her as a sort of sister or soul mate—but their steady flow of letters and their habit of taking long carriage rides along the Hudson River suggests the crushy intensity of youth, which only intensified with their separation during the summer of 1890, when Dashiell left New York for a cross-country train journey.[1]

She returned in late August complaining of only one small injury. Her right hand had been caught in the seat lever of her Pullman rail car and was now swollen and discolored.[2] She couldn't sleep with the pain. Finally, Johnny Rock's family suggested New

York Hospital,[3] where Bessie would be examined by a twenty-eight-year-old bone specialist and surgeon freshly released from Harvard Medical School, Dr. William Coley.[4]

Coley was a rising star in the surgery department, a skilled and caring clinician with a youthful enthusiasm for new ideas, such as germ theory and Joseph Lister's latest advances for controlling infection through sterilization techniques and vigorous handwashing.[5] These modern notions rendered surgery far more survivable for patients; they might have also put the young surgeon in a state of heightened awareness of both the astonishing invisible microbal world around him and the promise of further scientific advances on the horizon. Coley considered that he had entered medicine "at the most opportune time in a thousand years."

The young surgical intern examined Bessie Dashiell's hand.[6] He noted a slight swelling "half the size of an olive," like an extra knuckle where her metacarpal met the pinkie. He thumbed the mass; it did not move, but it was tender, and the girl winced. Coley carefully palpated Bessie's jawline and armpits and found them unremarkable. Her lymph glands weren't swollen. That suggested that the problem wasn't an infection; there was no immune response.

As a bone specialist and a surgeon, Coley's best guess was that her pain and swelling resulted from inflammation of the sheath-like sac that covered the bone of her little finger. To be certain, he needed to cut. Coley took his scalpel and drew a line down the girl's finger, parting flesh and membrane down to the bone. He noted that he did not find the great reservoir of pus he would expect from infection, and that the membrane was hard and gray. His diagnosis was periostitis, a subacute bone ailment. Dr. William T. Bull—his mentor and a legendary surgeon known as the Dapper Dan of the operating theater—agreed, and the young woman was sent home to let time heal this wound. But over the following weeks, Bessie Dashiell's Pullman pinch continued to worsen. And that didn't make sense. If all the symptoms resulted from the initial insult to the bone, they shouldn't have been getting worse.

Coley performed a second exploratory surgery on Dashiell, scraping more of the tough gray matter from the bone. But the swelling and pain continued to increase, and Dashiell began to lose sensation in one finger, then others. Now the young surgeon had to consider a more dire diagnosis and yet another surgery. This time Coley cut a slab of the gritty gray matter from Bessie's finger to be analyzed. A few days later, a report messengered from a New York Cancer Hospital pathologist confirmed his suspicions: Under the microscope, the "granular" gray stuff Coley had been scraping off Bessie Dashiell's bone was revealed as cancer. Specifically it was a sarcoma, and it was spreading. What little feeling Bessie had left in her fingers now radiated as pain. Coley prescribed morphine.

Sarcoma was a relatively rare form of cancer, a disease that affects the connecting tissues of the body such as the tendons and joints and ligaments. (It's distinct from what is commonly called carcinoma, which affects essentially everything else.) Treatment options for cancer, especially those of the bone, were extremely limited in 1890.[7] The only means the surgeon knew to get rid of the cancer was to cut off the hand itself.

Coley hoped to cut beyond the clean margins of the disease, while leaving the girl some useful length of arm. But the cancer had already spread. What had started in her pinkie now proliferated grotesquely through the landscape of her young body. Small buckshot-like nodes began to appear in one breast, then in the other. Soon, they were in her liver, and Coley was able to feel a large solid mass growing above the young girl's womb; perversely, he described it as being the size of "a child's head."[8]

Bessie Dashiell's decline was shockingly rapid. By December the young woman's porcelain skin pushed out everywhere in hard lumps. Her liver was enlarged, her heart was failing, and she was skeletally thin, surviving only on brandy and opium. The frail, drug-addled creature was almost unrecognizable as the pretty, plucky young woman who had walked into his offices only two months before, fresh from a cross-country adventure. There was nothing for the young surgeon to do but bear witness and provide the comfort of opiates. Dashiell died at home on the morning of January 23, 1891; Coley was at her bedside.

Coley would later admit that her death had been for him "quite a shock." Partly it was her youth, and his—he was new at the job, and only ten years older than Dashiell. And partly it was the rapidity of this disease, and his helpless flailing in the face of it. Perhaps his surgeries had even hastened the disease by scraping it loose into her bloodstream. Maybe he had made her suffering worse by trying to save her.

Despite his modern surgical refinements and degrees, Coley had offered Bessie Dashiell little that wasn't available in the blood-slick chairs of street-side barber surgeons or the numbing comforts of the barroom. He was determined to find a better way. Technological advances were rapidly ushering in the new century; each morning paper seemed to herald a stunning new scientific advance. In the previous ten years Karl Benz had invented a gasoline-powered engine, Charles Parsons had invented a steam turbine, and George Eastman developed plastic photographic film. Less than a mile from Coley's medical offices, Nikola Tesla and Thomas Edison were in a furious competition to build power stations capable of illuminating whole city blocks. It felt as though the whole world would soon be lit and the shadows of ignorance banished.

The records of everyone who had ever walked or limped or been carried through the hospital's doors were written in copperplate longhand in oversized ledgers. Coley turned the heavy pages, scanning the progress notes on every patient who had presented with a disease resembling Bessie Dashiell's. This was tedious work; the records ran chronologically, page after page, book after book. By immersing himself in the collective experience of cancer patients and their presentations, Coley thought, he might better understand the fatal course of Bessie's cancer. And if he was lucky, he might find an exception to it.

Seven years deep in the patient logs, Coley's attention caught on an unusual case history. It belonged to a thirty-one-year-old patient named Fred Stein, a German immigrant and house painter. He had arrived at New York Hospital in the winter of 1885 with a disfiguring egg-sized mass bulging from his left cheek near the neckline.[9] This was far larger than the one Bessie Dashiell had on her hand, but it was the same sarcoma.

Dr. William T. Bull, the head surgeon at New York Cancer Hospital, had operated on Stein to remove the mass.[10] When it came roaring back, Bull operated again. Yet again it reappeared and grew until it was as big as a man's fist. Bull performed a total of five operations on the man over the course of three years. It was impossible to remove all the tumor, and the case was considered "helpless." Skin grafts were attempted but were unsuccessful, leaving an open wound, which soon became infected with erysipelas.

Erysipelas was the name given to an infection caused by bacterium known as *Streptococcus pyogenes*, the bane of nineteenth-century hospitals. Under the microscope the bacterium appeared as little chains, like a bead necklace cut into small lengths.[11] On the ward, carried by the wind or bedding, these infection seeds infested open wounds and blossomed in the bloodstream. Infected patients broke out in fiery red rashes that started at the face and neck and spread rapidly, followed by raging fever, chills, inflammation, and, usually, death.[12]

Erysipelas was the most deadly postop killer in the nineteenth-century hospital and still ominously referred to as St. Anthony's Fire, as it had been since the Middle Ages.[13] The name referred to the speed of the infection's spread, its burn-like symptoms, and the desperation of the infected, who prayed for a miracle.

Fred Stein—deathly ill from an inoperable tumor, an open surgical wound on his neck, and infected by erysipelas—was assumed to be doomed. Instead, as the fire spread and Stein's fever raged, his surgeons noted something unusual. His tumor mass appeared to be melting away.

According to his hospital record, Stein survived the fever only to relapse again a few days later. He continued recovering and relapsing. Each time he slipped back into fever, his remaining tumor masses seemed to be withering and shrinking. Four and a half months later both the infection and the cancer were gone, and Stein walked out of the hospital. It was presumed that he had returned home to the immigrant slums of New York's Lower East Side, but his address was never recorded in the hospital records. That had been seven years earlier. What had become of Stein or his cancer since, nobody had bothered to find out. The only evidence of his existence and "miraculous" cure were the notes in the ledger.

Coley was intrigued. Here he had two patients that had presented with the same disease and been treated by the same methodologies at the same hospital, under the watch of the same physicians. And yet these patients had experienced wildly different outcomes. Dashiell had done well in surgery but died anyway. Stein had done poorly in surgery, become infected, and survived. It was so counterintuitive, it was tempting to look for causality. Had Stein survived *because* he'd become infected?

Either the observation on Stein was incorrect, or this incongruity offered a glimpse of something not yet understood. The only way to know more was to examine Fred Stein himself. And Fred Stein had last been seen walking out the stone gates of New York Hospital seven years before. Now, he could be anywhere, including underground. William Coley had landed upon a medical adventure. It would prove to be his strong suit.

Like many of his late-nineteenth-century contemporaries, Coley believed that the answer to the big questions of science were out there, somewhere, waiting to be discovered. The thinking wasn't so different from that of contemporary scientists using supercomputers to mine new insights from dumps of old data—except in the late nineteenth century, the answers were more likely to be uncovered via machete or microscope. That same year, radiation and X-rays had been discovered, and several new elements had been attached to the periodic table.[14] Fridtjof Nansen was attempting to reach the North Pole. Sir Richard Burton was bringing back tales of sea-sized lakes in the center of Africa. And now here was Coley, young and trained and ready. Coley wasn't one for the quiet sit-and-study routine of academic research; he had a quest to undertake.

Coley was a Connecticut Yankee from an old New England family, but he wasn't a complete stranger to the newer faces of 1890s' immigrant America. While still a student he'd worked aboard a brigantine on the rough Atlantic passage between the Azores and the wool mills of coastal Rhode Island and Massachusetts, and his wards at New York Hospital treated the huddled masses arriving from every corner of the globe. Many settled in the tenements in Manhattan's Lower East Side, a ghetto segregated from uptown society by the hard line of Fourteenth Street, but just south of the hospital.

After putting in a shift, Coley (now the personal surgeon to Rockefeller) took a hansom cab downtown, climbed out in his tailored English suit, and started walking the streets made famous to uptown slum tourists by photographer Jacob Riis's 1890 book *How the Other Half Lives.* Coley himself wrote little about his forays in search of Fred Stein, and so it's difficult to imagine whether it plays as comedy or drama. Probably it was both. It took weeks of combing the tenements, walking up and down stairs, knocking on doors, describing and gesturing. But finally, improbably, on a second-floor landing, a door opened to his knock and William Coley found himself face-to-face with the man himself.

A photograph of Fred Stein, provided in Coley's published report in the medical literature, shows a tall, gaunt man with the glandular severity of an Old Testament hermit. His hair was black with high, chopped bangs like a kid might do with safety scissors. High, polished cheekbones framed a goatee that extended like a black curtain from nose to collar, years of growth pulled and cut square. You must assume the mouth. Only the back of his hair was long, a cascading mullet that only partially covered the puckered scars of disease, surgery, and infection.

If Coley was surprised he did not say. The real surprise was that Stein was not only alive but apparently enjoying excellent health. After some initial awkwardness and comically pidgin German, the young doctor was able to persuade Stein to return with him to New York Hospital to be examined by his original physician, William

T. Bull. Bull confirmed this was the same Fred Stein on whose notes he had written the terminal prognosis and discharge, dated 1885.

Something had changed Stein's cancer, and with it his fate. The only observable *something* between Stein's failed cancer surgery and incredible cancer remission was the bacterial infection. If that infection had somehow cured a case of "undoubted sarcoma," Coley wrote later, "...*it seemed fair to presume that the same benign action would be exerted in a similar case if erysipelas could be artificially produced.*"[15]

And Coley couldn't wait to be the one to artificially produce it.

~

Coley's observation was astute and important, but hardly unique. Physicians had been describing spontaneous regressions of disease, including cancer, for thousands of years. Many had been observed to be coincidental to, or perhaps even in reaction to, the introduction of a new, different disease to the patient's system, including erysipelas. By the time Coley was making his observations about Stein's infection, such notions had become a regular feature of anecdotal medical hypotheses. Only two years before, Anton Chekhov, the Russian physician and playwright, had written about the apparently well-known phenomenon to a friend.

"Cancer is not a microbe," Chekhov began in an 1890 letter from Moscow to his colleague Alexei Suvorin. "It is a tissue growing in the wrong place and like a noxious weed smothering all the neighboring tissues...It was observed long ago that with the development of erysipelas, the growth of malignant tumors is temporarily checked."[16]

Over two hundred years earlier, Friedrich Hoffmann, in his 1675 medical treatise *Opera Omnia*—a modest work on the stated subject of literally "everything"—noted that an outbreak of St. Anthony's Fire had driven from his patients the other diseases already in situ, much as fire clears a diseased forest. French physicians Arsène-Hippolyte Vautier and S. L. Tanchou claimed to have effected hundreds of successes against breast cancer by means of infection induced by dressing their patient's wounds in soiled bandages worn previously by other infected patients. The sign that the desired infection had taken hold was a "laudable pus" that flowed like human sap from the wound.

Exceptional stories like these are found throughout medical history.[17] And for hundreds of years they remained just that—stories, anecdotally compelling and scientifically inexplicable. Nevertheless, they were enough to provoke occasional speculation and experiment. The result was an often ethically perilous version of mad scientist immunotherapy—human experiments without systematic methods, accountability, or follow-up. Most were performed on poor women—infecting breast cancer patients with gangrene, or adding a hypodermic full of syphilis to the wombs of women suffering from uterine cancer. (The latter was performed by a Belgian physician in 1851 and justified by the dubious claim that prostitutes were not known to suffer from uterine cancer.)[18]

In the 1890s these repeated medical observations of spontaneous remission of cancer attracted renewed international scientific interest.[19] In fact, at the very moment that Coley was formulating a plan to intentionally replicate Stein's accidental erysipelas infections, a physician named Friedrich Fehleisen had already begun.

Within a month of finding Stein, Coley had also found Fehleisen's data in a foreign medical journal. Fehleisen had identified the specific bacterium strain that caused erysipelas, had injected this bacterium in five patients, and was enthusiastic about the possibilities. Coley read this and was even more convinced that it was just such a postoperative bacterial infection that had rid Fred Stein of his terminal cancer. He apparently hadn't gotten the subsequent news that Fehleisen's experiment had cost several of those patients their lives, and Fehleisen his medical career.[20]) The only way to prove it was to reproduce a similar effect in another willing and desperate patient. He found exactly what he was looking for in an Italian immigrant we know only as "Mr. Zola."

∼

By the time Mr. Zola stepped off his migrant ship to the New York docks, a conspicuous morphine habit was the least of his problems; certainly it was his only relief. Zola had presented to Coley's hospital in March 1891 with a recurrent sarcoma in his neck that a previous surgeon in his native Rome had already operated on to remove.[21] The cancer had soon grown back and spread, and now another tumor "the size of a hen's egg" in Zola's throat prevented him from speaking, eating, or even swallowing. He had a hacking cough (likely the same cancer metastasized to his lung), and few options but to present himself to the charity ward of New York Hospital, where he was operated on by William Bull. Bull cut out a piece of the neck tumor, "about the size of an orange,"[22] but he couldn't get it all without killing the man. Bull concluded Zola to be a hopeless case; Coley estimated that Zola had at best a few weeks left to live. Apparently, Zola believed it too. It's difficult to imagine any other circumstance under which he would have willingly allowed himself to be infected by a deadly bacteria.

Erysipelas wasn't a disease anyone would foster intentionally. It thrived in the close quarters, poor ventilation, and inadequate bed linens of the poorer sick wards. Though both William Bull and Zola consented to the experiment, the risk was considered too great to conduct within their hospital. Zola would be infected at home.[23]

Coley wasn't a fastidious experimental data collector, but he was a well-trained physician, a gifted surgeon, and a keen observer. He was also persistent and fairly lucky. Modern clinical trials of any drug have a standardized protocol to ensure that they are reproducible and correlate causes to effects. Coley basically winged it. His experiment was less a clinical trial than the monkey-wrenching of an intuitive biological mechanic. He wanted to cure Zola's cancer, not write a paper. The papers would come later.

And so, in the course of testing the bacterium on Zola, Coley switched between two different bacterial strains prepared by two different sources, which he administered in two different ways. At first, he made small cuts in his patient and applied the bacterium, which he had grown on a gelatin, directly into the incisions, but he soon found this method unsuitable and abandoned it mid-experiment. He then cultured other samples of the bacterium in beef broth and injected anywhere from half a gram to two grams away from the incisions, but neither of these did much more than to give Zola a slight fever, a quickened pulse, and light chills—nothing like the symptoms of the dreaded St. Anthony's Fire Fred Stein had endured.

Finally, Coley determined that the problem might lie with the virulence of his particular strain of bacteria, and he asked two colleagues from Columbia University's College of Physicians and Surgeons to mix up a stronger brew. This he injected in large doses directly into Zola's unhealed neck wound and various other places on his skin. Within a few hours Zola's body responded with localized redness.[24] Zola still couldn't speak because of the tumor blocking his throat, but he could wince and hold his aching head. The chills and vomiting spoke for themselves, but at 101 degrees F, his fever was only 0.5 degrees higher than it had been using Coley's first bacterial batch.

Still, Coley believed the treatment was working and continued plugging away at his patient. After a month of steady injections, the tumors in Zola's neck and throat seemed "diminished"—sometimes "appreciably diminished"—in size.[25] That was good, but it wasn't exactly the spontaneous remission he'd read about with Stein. Undeterred, Coley determined to push harder, and with stronger toxins.

That summer of 1891, Coley decided to forgo even a brief vacation to remain in the city, injecting bacterial toxins into his patient. Meanwhile his hospital colleague Farquhar Ferguson would be spending his holiday on a miniature grand tour, sampling the culture of the European continent. Coley asked Ferguson to bring back a souvenir; he wanted a bit of deadly infection, fresh from Berlin.

As Debra Jan Bibel points out in her comprehensive 1988 book, *Milestones in Immunology: A Historical Exploration*, our view of the world is often shaped by the lenses through which we examine it. In the late nineteenth century that biological view was defined by the literal lens of new and powerful microscopes, and the startling bacterial world that technology had suddenly rendered visible.

Suddenly, the factors responsible for illness, infection, and beer were realized to be living creatures. Different types of bacteria, it was believed, made different types of poison, or toxin; the body's curative response was with some sort of antitoxin (or, as it would be called later, an antibody) to cancel it out.[26]*

In this bacterial age, Robert Koch was practically a household

---

* See Appendix C for more information.

name. Koch was a prodigious collector of deadly toxins, most famous for isolating the deadly bacteria that caused anthrax in his Berlin laboratory. If anyone could provide Coley with a whoppingly deadly dose of erysipelas, it would be him.

Ferguson arrived back in New York at the beginning of October, his unusual travel souvenir from Koch's laboratory in carefully wrapped glass vials. Koch hadn't disappointed; his erysipelas sample had been collected directly from a corpse only days before Ferguson's visit. This was the good stuff, potent and fresh. Coley wasted no time. On October 8, he traveled back to Zola's Lower East Side rooms, loaded up his syringe with five decigrams of the new German bacterial toxin, and injected the toxin directly into the tumor in Zola's neck.

This was definitely the good stuff. Zola's internal temperature started to climb; within the hour it would hit 105 degrees F. Meanwhile the infection boiling beneath Zola's skin at the injection site darkened and spread across his upper extremities like fire consuming paper.

Zola was pushing the boundaries of physical endurance, but by the second day of fever, the sweating, trembling patient finally produced the results Coley had been hoping for. Zola's tumor seemed to be physically "breaking down." Soon it was melting down his neck like some horrible ice cream cone. "A discharge of broken-down tumor tissues continued until the end of the attack," Coley wrote. At the end of two weeks, Coley reported, "the tumor of the neck had disappeared."

The tumor on Zola's tonsil was still there, but it had shrunk enough that Zola could eat again, and the patient "gained rapidly in flesh and strength." Soon Zola felt well enough to leave bed and get back to his business, which, Coley noted in his final sentence on the man, included "a confirmed morphine habit which he had contracted previous to the inoculations."

Coley examined Zola two years later, and again after five years, and found the man still in good health. (Soon after, he returned to his native Italy, where he died, eight and a half years after his injection regimen, causes unknown.) What Coley witnessed with Zola was not a typical reaction; in fact, the success with that particular bacterial "toxin" has never been fully explained.[27] But *something* had happened, and that something wasn't magic.

The gap between the observations of so-called spontaneous cancer remission following infection and the scientific understanding of the complex, microscopic, and yet-unguessed-at immune biology responsible for it would be the bane of cancer immunotherapy researchers for a hundred years. Here was a field where, time and again, experiment and observation outpaced even the faintest understanding of the unimaginable complexities of either the immune system or cancer. As a result, cancer immunotherapy retained a certain naturalist's air about it, a field of both science and stories—observations of therapies that worked for some people and not others, results that were confoundingly difficult to reproduce, immune responses that cured cancer in a mouse or in a petri dish but did nothing for humans, all of it scientifically mysterious. As Stephen S. Hall put it in his 1997 immunology masterwork, *A Commotion in the Blood*, "The tyranny of the anecdote, the boon and bane of immunotherapeutic interventions, had formally begun."[28]

∾

Zola had been a one-off, too unstandardized and uncertain to qualify as a properly performed scientific study, or as proof of anything. Now Coley attempted to reproduce his success, patient after patient, method after method. By now, his work with deadly bacteria had moved him uptown to 106th Street and Central Park West and the ventilated gothic towers of the New York Cancer Hospital[29] (which would later be renamed General Memorial Hospital; today we know the entity as Memorial Sloan Kettering Cancer Center, or MSKCC). Coley tried direct injection; he tried rubbing the bacteria in; he tried scarification techniques, combinations, and repetitions. Over the course of three intense years, Coley gave repeat inoculations to twelve patients presenting with a variety of cancers. He had more failures than successes.[30] He triggered the desired fever reaction[31] in four of his patients, and that plus a positive tumor response in another four (including Zola). All of the patients who showed a response had sarcomas. Four of his patients died, two as a result of the bacterial infection Coley had initiated. Coley couldn't predict who would react to the bacterial toxins or what dose to give, or by extension, whom he might help and whom he might accidentally kill. It was an intolerable situation, not to mention a dangerous and unethical one. It endangered his own medical practice as much as it endangered his patients.[32]

Infecting his patients with the live bacterium was too risky, but it wasn't the whole living microscopic animal he was after anyway, just the "toxic products" that he believed destroyed tumors. Now Coley started working on a plan for "isolating and using the active principle of the germ."[33]

The idea was based on the serum-centric view of contemporary biologists, and the fact that inoculating a patient with a dead or an inactive form of a bacteria was the fundament of vaccination.

That summer, an especially deadly strain of the bacteria was grown in the laboratory. The live bacteria were overheated and killed,[34] then the broth was strained through a porcelain filter to remove the dead bodies of the bacteria themselves. The ruby-colored juice that flowed through the other end of the filter was assumed to be just the "toxins" from the bacteria. This had to be the stuff. Coley injected this new brew into a fresh group of terminal sarcoma patients. The serum had some of the desired effect—light fever, rash, and chills—but not enough.

Now Coley was in a bind. He needed to find the sweet spot between too little toxin and too much. Once again, Coley got lucky. Just as he was asking the question, a French medical journal happened to include a new study that supplied the exact answer.[35]

The study showed that the erysipelas bacteria Coley was using became far more virulent, and produced a far more powerful toxin, when cultured in the same incubator together with another bacterium strain, called *Bacillus prodigiosus*.[36] With this recipe Coley hoped he'd finally found a compromise between a deadly and an ineffective toxin. In fact, he'd stumbled onto the perfect combination of bacterium, which produced a synergistic toxic effect.

As its name suggested, *B. prodigiosus* proved to be a very prodigious little bacillus indeed, and produced a toxin with a unique effect on the human immune system (ones that are being looked at today as cancer therapies, some in clinical trials).[37] Now he needed to find a subject on whom to test his potent new bacterium-combo toxin.

Coley would finally get his chance in 1893, with a sixteen-year-old boy whose belly seemed pregnant with an eggplant-sized sarcoma. John Ficken, like most of Coley's subjects, was a patient with nothing left to lose. The massive tumor had invaded the wall of his abdomen, as well as his pelvis and bladder; biopsy suggested that it was malignant.

Coley started Ficken off gently, with a low dose of his new

toxins. When Ficken didn't respond he titrated up, first half a cc, then more, every couple of days. Finally, the boy had the classic reaction Coley had witnessed with his previous toxins—St. Anthony's Fire.

The treatments started on January 24 and lasted ten weeks. By the time Coley stopped the injections on May 13, the tumor mass had shrunk by 80 percent. A month later it was no longer visible to the naked eye, but it could still be palpated. Coley sent the boy home a few weeks later. Ficken was feeling well, looked normal, and, despite the loss of the tumor, had gained weight.

Eventually, of course, Ficken died—suffering a heart attack on a subway car outside Grand Central Station. At the time he was forty-seven years old. Coley's bacterial concoction—later patented as Coley's Toxins—had cured his cancer, at least for the additional thirty-one years of his lifetime.

~

Coley published in the usual medical journals, but excited as he was, or perhaps impatient, in 1895 he wrote up his own volume on his sarcoma treatment and brought it to the offices of the Trow Directory, Printing and Bookbinding Company on East Twelfth Street. The volume was part academic medical journal, part testimonial, and was the same dimension as a religious pamphlet or museum guide. (The size is still a sort of unofficial standard for some resident quick guides that fit perfectly into a pocket of a medical resident's white coat.)

"I am conscious that the treatment of inoperable tumors is a very trite subject," Coley began his treatise, "yet in consideration of the fact that practically no advance has been made in this field since the disease was first known, I am sure I need offer no apology if I can show that there has been even a single step forward."[38]

In fact, Coley was certain he'd made not a step but a leap.

"My results in thirty-five cases of inoperable tumors treated with the toxins during the past three years, were reported in detail at the last meeting of the American Surgical Association at Washington, May 31st, 1894," Coley wrote, "and will be only briefly referred to here." At which point Coley shared his proprietary recipe for the toxic home brew.

The recipe called for a pound of meat, lean and chopped and left overnight in 1000 cc's (about a quart) of water. In the morning, remove the meat; the remains were the raw beginnings of a boullion broth. This meat-water should then be cloth filtered, boiled, and filtered again. Season with salt and peptone (a partially digested protein, enzymatically broken down into shorter amino acids so that it can be digested by simple bacteria; essentially microbe chow). Another pass through the straining cloth and another boil will result in a clear consommé. Finally add the deadly bacterium, and you're ready to serve mankind.

∿

At least fifteen versions of Coley's toxins were in use during Coley's lifetime. (Parke Davis manufactured the most widely used commercial version; the Mayo Clinic made another for their patients, and continued well after others had quit the field.) Coley had indeed invented a sometimes effective cancer cure by intentional immunotherapy, not that he realized it as such at the time.[39] What Coley achieved could have been a breakthrough, if the results had led to systematic further investigation of the phenomenon, and a push into the basic science behind it. Instead, the opposite happened;

Coley's results were a century ahead of any science that might make sense of them, and they were largely construed as quackery.

Coley had theories about the agents at work. But he had no real understanding of the immune system, or the nature of cancer, nor any inkling of the genes, mutations, antigens, or any of the biology necessary to bridge the gap between what he'd observed and something approaching bench science. The mechanisms by which immune cells recognize disease had not been discovered; not even the immune cells themselves had been discovered. Nonetheless, over the course of the following forty years Coley continued to treat hundreds of patients with his toxins.

More recent scientific assessments of the effectiveness of those treatments vary—one by Coley's daughter reviewed more than one thousand of Coley's patient records and reported finding some five hundred cases of remission; a controlled study in the 1960s reported Zola-like results in twenty of ninety-three patients.[40] The numbers vary wildly and much of the methodology is questionable, but when reading all the academic analyses and reviewing the more recent experimentation, the takeaway is always the same. Coley was no quack.

Careful regulation of a patient's fever—a laborious and personal process—seemed key to his successes. That factor, and the enormous variability in the exact formulations and strengths of the toxins available to other physicians, made Coley's results difficult to duplicate. But that doesn't change the general consensus that in Coley's hands, his toxins did sometimes work, and they sometimes worked well.[41] And the reason they worked is now believed to be that they somehow triggered an immune response, or unleashed a previously blocked one.

But as a medicine, Coley's Toxins would not last.[42] Parke Davis stopped production in 1952. By 1963, the Food and Drug Administration would no longer acknowledge Coley's Toxins as a proven cancer therapy.[43] The fatal blow to Coley's vision of immunotherapy occurred two years later, when the American Cancer Society put the fluids on its list of "Unproven Methods of Cancer Management." They might have well have called it "the quack list."

Ten years later the ACS would reverse itself and remove Coley's Toxins from this list of shame, but the damage was done.[44] The retraction garnered less notice than the original infamy. His name, if it was known at all, was associated with the absurd miracle medical claims of the gaslight era of radiation gargles and patent medicines; whatever glimpse at the possible interaction between the immune system and cancer he had provided now seemed like a delusion or fraud.

Ideas can be powerful viral things that spread like a forest fire. They can also be snuffed out like candles. It takes only a generation for an idea to be forgotten. And a generation of researchers, scientists, and doctors went through training hearing nothing about Coley or a successful if still mysterious illustration of the potential of the immune system to be induced to interact with and protect against cancer. For thirty years Coley and his methods were virtually unknown to oncologists, and, as Hall writes, those who did "lumped them together with such controversial therapies as Krebiozen, laetrile, mistletoe and orgone boxes."[45] Good oncologists looked to the more modern and promising scientific therapies like radiation and chemotherapy. And when those physicians trained the next generation, they taught them to do the same. If you were smart and science-minded and came of age in the '80s or '90s, you weren't trained to put stock in Coley.

Coley's legacy might have died with him had it not been for the efforts of his daughter, Helen. Helen Coley Nauts had traveled with her father on many of his speaking engagements, had seen him rise to become a wealthy and famous man, and had seen his fall. Nauts understood her father's work even as he had not, and in doing so she helped carry his ideas into the present generation.

Near the end, Helen had watched her father at conferences fending off attacks against both his data and his person. The most vigorous of those attacks came from Memorial Sloan Kettering, the cancer center Coley had helped establish, and where his approach to cancer had first been supplanted by radiation therapy, which was viewed as more modern, with more quantifiably scientific results. It also didn't hurt that, although the radium required to make radiation therapy was considered one of the scarcest resources on earth, the chief benefactor of the hospital at the time was a mine owner supplying Memorial and its charismatic and powerful president, Dr. James Ewing, with as much as it needed. Memorial's hoard of eight grams was reported to include the original supplies of Marie Curie, and represented most of the known radium supply on the planet.

Together, Ewing and Coley had turned Memorial into the world's first cancer research center.[46] Now Ewing was Coley's boss, and his leading critic. He publicly denounced Coley's Toxins as a fraud and a sales scheme. Soon, every patient who came to Memorial with a bone complaint would get a full dose of Ewing's exclusive radiation therapy. The results, of course, were disastrous. The mortality rate was 100 percent.

Coley requested a five-year trial of his toxin vaccine, as it was then considered, in order to evaluate its effectiveness against bone cancers such as sarcoma. Coley claimed to have no statistical data to prove that his treatment was effective, but then, neither did the advocates of radiation therapy and amputation. What

Coley did have were survivors. The advocates of radiation therapy had none.

Coley never did get his five-year trial; he died within a year of asking for it. But his daughter never forgot it. In 1938 she went to the family country estate in Sharon, Connecticut, and discovered all her father's papers—some fifteen thousand of them—bundled and stored in a barn at the edge of the property. It wasn't that Coley didn't have data; he just didn't have it organized.

Working tirelessly (and funded in part by a small grant from Nelson Rockefeller, the son and heir of her father's patron and Bessie Dashiell's soul mate, John D. Rockefeller Jr.) Nauts organized her father's heap of observations, correspondence, and notes into something more organized and academic. With a high school education but a lifetime of tutelage under a master physician and thousands of hours of careful study, Nauts set out to convince anyone who might listen that her father's approach to "the use of bacterial products in malignant disease" was, at the very least, worth a more serious investigation.

~

William Coley had believed that the "toxins" from his bacteria acted as a sort of poison against cancer—a natural chemotherapy. By the 1940s, following the death of Ewing, Memorial had transitioned from being a "radium hospital" to one that used chemical poisons—chemotherapy—as cancer-killing therapy. Nauts hoped to pursue her father's work with the new director of the hospital, the eminent physician Dr. Cornelius Rhoads. During World War II he had served as the chief of research for the Chemical Wartime Service of the US military forces, the same group that had discovered the potential for mustard gas as an agent of chemotherapy against cancer. Rhoads would become chemotherapy's biggest booster, and he helped usher in a modality for cancer treatment that continues to be the norm. But Rhoads wasn't interested in Coley's toxins either.

Nauts had no formal training in medicine; she couldn't explain why her father's medicine had worked. But she had his data, and she had a theory as to the mechanism behind it.

Coley's Toxins, she posited, weren't toxins at all. They were a stimulant. They didn't act on the tumors directly; they somehow worked "by a stimulus to the reticulo-endothelial system."[47] The system she was referring to is what we now call the immune system. She was broadly correct. Rhoads still wasn't interested.[48]

Finally, in 1953 Nauts once again made an appeal to Nelson Rockefeller, the son of her father's former benefactor. His father's friendship and heartbreak over losing his "adopted sister," Bessie Dashiell, had inspired him to a life of cancer-focused philanthropy, supporting William Coley's research, creating Rockefeller University, and helping Coley and Ewing fund the country's first cancer hospital. Now the younger Rockefeller gave Nauts a grant of $2,000, with which she and her associate, Oliver R. Grace Sr., founded an organization Nauts hoped might keep her father's ideas alive, and fund others in similar pursuit. That organization, the Cancer Research Institute, had its offices on Broadway in lower Manhattan. It still does.

CRI was the first institute dedicated exclusively to promoting the ideas of cancer immunotherapy. For many years their phone did not ring.

Chapter Three

Glimmers in the Darkness

Blut ist ein ganz besonderer Saft. (Blood is a very special juice.)

—GOETHE

In retrospect, it's surprising what we *didn't* know about our bodies, and how recently we didn't know it. We had a pretty clear picture of the planets in the solar system and the composition of moon rocks before we had a working understanding of what was happening in our own bloodstreams.

The study of immune biology started with the microscope, and a mess of cells strained out of the blood by a biologist's porcelain filter. The ones that were red were recognized as the blood cells that shuttle oxygen through the body. The cells that weren't red were called "white," in the sense that non-red wine is called white. These white blood cells are also called *leukocytes*. (The Greek root for "white" is *leuk* and for "cell" is *cyt.)* The term still refers to any cell that's part of the immune system.

Immune cells were originally assumed to all be the same. It would take more than a simple microscope, however, to reveal that our bloodstream in fact contains an exotic ecosystem of specialized players bound in an elegant and potent web of personal defense.

~

The first aspect of immune response to be grasped by nineteenth-century biologists was the oldest and most primitive, a 500-million-year-old personal defense system we call the "innate" immune system.[1]

The innate immune system is charismatic and deceptively straightforward. It also happens to have cells big enough to be seen wiggling and eating under the microscope. That includes amoeba-like cells adept at squeezing between body cells and patrolling our perimeter (inside and out, we have a surface larger than a doubles tennis court), looking for what shouldn't be there and killing it.

These cells include small blobby smart patrollers called *dendrites* (remember them for later) and similar but larger blobby characters called *macrophages* (literally, "big eaters"). Among their other jobs, these serve as the garbagemen of the immune system. Mostly what they eat are retired body cells—normal cells that have hit their expiration date and politely self-destructed. They also eat bad guys.

Macrophages have an innate ability to recognize simple invaders. These foreign, or non-self cells, are recognizable as foreign because they look different—that is, the fingerprint of chemical arrangements of proteins on their surfaces is different. Macrophages look for anything they recognize as foreign, then grab and gobble it.

These cells also end up saving little pieces of the invader cells they kill, creating a show-and-tell for the rest of the immune system. (We've also recently discovered that some innate immune cells are more than just simple eaters and killers—they seem to be the brains of the larger immune system.)

Innate immune cells are tuned to recognize the usual suspects—the bacteria, viruses, fungi, and parasites that evolved right alongside us and account for most of what needs defending against.

Where there's one invader there are probably more, so the cells of the innate immune system can also call for local reinforcements. The call for help is chemical, in the form of the hormone-like proteins called *cytokines*. Many cytokines are like a distress beacon, with limited range and longevity, to prevent overreaction. There are many different flavors of cytokines, conveying many different messages. Each begins a complex choreography of chain-reaction defense responses within the body.

The result is a surprisingly sophisticated chemical communication that can call for more blood supply and tell the small blood vesicles (capillaries) to become more leaky, so fluid and reinforcements can flood in between the gaps (what we know as inflammation), and even stimulate the local nerves to send out extra *ouch!* signals (so you pay more attention to the problem—and maybe remember not to do it again).

That's what an immune system looks like for nearly all life on earth. It works fine for recognizing and killing the usual suspects of disease, providing a rough and ready response effective enough to mop up most invading threats in just a few days.

But more recently evolved critters on the tree of life—vertebrates with jaws, like us—also have an additional type of immune army, one capable of adapting to meet new challengers. This is the "adaptive" immune system, and it is able to face, fight, and remember *unusual* suspects: invaders that the body has never encountered before.

The major players of this adaptive immune system are two distinct types of cells that travel our bloodstream, with distinct tools of defense.[2] These are the B and T cells.

Diseases evolve and adapt. Nature invents new ones all the time. The B and T cells are part of a system that adapts to counter them. In terms of attacking cancer specifically, it's the T cells we care about. But both B and T cells play a role in the cancer immunotherapy story.

∾

Vaccines are the most successful form of immunotherapy, one we've been familiar with for hundreds of years. Their biological mechanism is dependent on the adaptive immune system.

Vaccines train the cells of the immune system on a harmless sample of a disease that it might encounter later. The introduction allows the immune system to build up forces against anything that looks like that sample. Then later, if the live disease does show up, an immune army will be waiting for it.[3]

Both B and T cells are involved with creating immunity. B cells were discovered first, so they get first billing.

Before these immune cells migrate into our bloodstream, they mature from stem cells in the marrow of our bones. B cells[4] have a unique method of defending us against the stuff that causes disease. They don't kill disease cells directly. Rather, they are factories that spit out antibodies—sticky, Y-shaped molecules that grab and hold on to foreign or non-self cells, and mark them for death.

Antibodies were originally called antitoxins, because they were assumed to be the stuff in blood that neutralized toxins—little customized antidotes that matched the poisonous molecules of disease like a lock to a key, canceling them out one by one.

B cells (and T cells) need to be ready to recognize anything that is non-self. This is possible because non-self, foreign, or diseased cells look different from normal body cells, at least to the discerning immune system. The difference is superficial—the outside of the cell is different. Foreign or sick cells have foreign proteins on their surface. The molecular marking is a distinctive bad cell fingerprint. These telltale fingerprint arrangements of foreign proteins on the surface of non-self cells are called *antigens*.

B cells create antibodies capable of recognizing the antigen fingerprints for even unknowable threats through an ingenious random genetic mix-and-match process that allows for 100 million different antibody variations. This variety is enough to ensure that at least one will match up with one of the many millions of possible protein arrangements of a foreign antigen. Every B cell makes antibodies to fit a randomly assigned antigen type. It's something like a lottery ticket approach to recognizing random strangers. Every potential combination is covered by one or another of the B cells. It really only takes one antibody recognizing a foreign antigen to kick off the immune response.

Here's how that works:

There are an estimated 3 billion B cells riding around in your bloodstream, each covered with sticky antibodies designed to match up with the antigens of diseases it will probably never meet, and which may not even exist.[5] B cells spend most of their short lives floating around until they happen to get lucky and come across the corresponding unique antigen of a pathogen (such as an unfamiliar bacteria, virus, fungus, or parasite).

If the antigen they encounter happens to match up exactly with the unique antigen receptors of a particular B cell's antibodies (which stud the B cell surface like cloves on a Christmas ham), that B cell snaps into action, producing clones of itself, identical daughter cells all born with the same "right" antibody.

In twelve hours, that B cell can make twenty thousand cloned copies of itself, and the process continues for a week. Each new-made member of the B cell clone army also becomes a new factory, producing just that antibody against that disease cell.

Now it's time to attack. The antibodies on the B cell surface fly out like sticky guided missiles at a rate of two thousand per second. Each of these antibody missiles has only one target: the unique non-self antigens on those foreign cells. They can see nothing else. The antibodies find and stick, accumulating on their target like burrs on a dog. Not only does this trip up the disease cell, it also acts like a blinking neon sign that catches the attention of the wandering blob-like macrophages, drawing them toward a free foreign meal. The antibodies are sticky to the macrophages too. They bind them to their dinner. They also seem to stimulate the appetites of "nature's little garbagemen" (a process known as *opsonizing*, from the German word that means "prepare for eating"). The foreign invader cell gets stuck, then gobbled.

It's a fantastically elegant and sophisticated defense that ramps up a response to a new disease in about a week. When the threat is over, most of the B cell army dies off, but a small regiment sticks around, remembering what happened, ready to snap back into action if the threat shows up again.

That's called immunity.

B and T cells look nearly identical under an optical microscope (part of the reason that, for most of the twentieth century, there was no such thing as a T cell). Just like B cells, T cells recognize a foreign antigen and ramp up a clone army to attack it. But T cells recognize and kill sick cells in a totally different manner.

~

Eventually it became clear to biologists that all those white blood cells that looked so similar under the microscope didn't look *exactly* the same, or function in the exact same way. By the 1950s it had been observed that some of the small lymphocytes (immune cells) also traveled differently through the human body.

B cells were known to originate in the bone marrow, travel the bloodstream for a while, and die. But some of these B-like cells seemed to take an extra side trip into a mysterious butterfly-shaped gland located just behind the sternum in humans, called the thymus; more of these cells were observed pouring back out of the thymus into the bloodstream. Even stranger, more came out than went in. Their numbers were sufficient to replenish the whole stock of B cells four times over, and yet, the overall number of lymphocytes in the body seemed to remain constant. So where did they go? The mystery of the disappearing lymphocytes was only cracked in 1968, when an experiment was able to follow them and find that the odd B-like cells that dumped into the bloodstream from the thymus were the same ones that later cycled back through the thymus. And many that went in never returned. It was as if they were being made, recycled, and perhaps modified in this strange gland.[6]

Experiments demonstrated that lymphocytes that cycled through the thymus were in fact very different from the familiar B cells. These cells seemed uniquely responsible for very specific aspects of immune response, such as organ rejection after surgical transplant.

The biological model that had all lymphocytes as B cells originating in the bone marrow didn't match the new observations. Which begged the question: Was there a different type of lymphocyte, one that came from the thymus instead? A white blood cell involved with adaptive immunity that wasn't a B cell? And if so, what should they call this thymus-born cell?

It was a surprisingly contentious question. When a young

researcher named J.F.A.P. Miller proposed to his colleagues at a 1968 immunology conference that perhaps they should consider that there were *two distinct* types of lymphocytes—B cells from bone marrow that made antibodies, and T cells from the thymus that somehow worked differently—he was publicly reminded that *B* and *T* are the first and last letters of *bullshit*.[7]

But, of course, Miller was right, and by 1970 it was generally accepted that these T cell lymphocytes, or "T cells," were different from the B cells that made antibodies.

It would be another five years before the picture was further complicated—or clarified, depending on your perspective—by the important realization that there were also several distinct types of T cells.

Immunologists distinguished two of the main ones with typical flair as "CD8" and "CD4," but they're better known as killers and helpers.[8] *Killer T cells* are the single-minded bruisers of the immune team, while *helper T cells* serve as a sort of quarterback for that team, "helping" coordinate the larger immune defense game plan by broadcasting a complex array of chemical signals, or cytokines.[9]

Finally, the bigger immune picture was starting to make sense. The T cells had been a missing piece. Their discovery provided a workable explanation for most of what had been observed about our reaction to illness and disease.

It goes like this.

The cells of the innate immune system respond quickly to familiar invaders, the usual suspects. Usually they are sufficient to do the job. Sometimes they just hold off the invaders while calling for reinforcements. But sometimes the invaders are unfamiliar, and an adaptive response is necessary.

Meanwhile, the B and T cells of the adaptive immune system

have started ramping up a response by making billions of copies of themselves, a clone army of the version of the lucky cell that happened to recognize the foreign antigen. That takes five to seven days.

Sometimes the defense is tag-team. The B cell antibodies gum up bad guys such as bacteria and viruses that make their way past the skin and mucosa layers of the epidermis and into the bloodstream, something like Spider-Man webbing up villains, so they can be collected later. They bag and tag. Then the macrophages gobble them.

But the B cells can't always stop all invaders in time. Sometimes the agents of disease get in, overwhelm the defenses, and infect a body cell.

Viruses inject body cells with their virus DNA. Once that's inside the cell, it's too late for the B cell to stop it with antibodies. Eventually, that infected body cell will become a factory for more viruses, cranking out reinforcements for the disease. To prevent that and safeguard the body, that infected cell needs to be killed.

If a virus does make it into a normal body cell and infects it, that cell changes. It starts expressing different proteins on its surface; it looks different, foreign.

It's up to the T cells to recognize those new foreign antigens of a self cell gone wrong and to kill that cell, up close and personal. Recognizing a sick body cell, locking in on the telltale foreign antigen, and killing that sick cell are the T cell's specialty.

After the attackers are defeated, most of the immune clone army dies off, but a few remain and remember. If that attacker shows up again, it won't take a week to clone up a new army and mount a defense. The body is ready.

And *that* is immunity.

This wasn't a complete picture (and of course it's far more sophisticated and interesting, and still being discovered—an entire exotic coral reef ecosystem that is described here as a goldfish bowl). But for scientists trying to figure out how the immune system did its thing, this new B and T cell model matched what they were seeing in almost every disease—with one horrible, glaring exception.

∽

Cancer was different. It was a sick body cell, no longer a self cell. But it wasn't infected—it was mutated. It was a disease that T cells didn't seem to recognize.

Most scientists believed the reason was that cancer cells were too similar to normal self cells for the immune system to recognize as foreign. That belief about the immune system and cancer was held by most cancer researchers, most oncologists, and most immunologists, and it corresponded pretty neatly to most observations about the disease. The immune system didn't attack it. You didn't feel sick until the cancer's unchecked growth crowded out your vital organs. Until then, there were none of the usual symptoms of fighting off a bug—no fever, no inflammation, not even a runny nose. That was the rule, and there were no exceptions.

Which meant the idea that you could help the immune system do its natural job and recognize and kill cancer cells was one that would never work.

∽

The scientific consensus on this point was fairly complete, and tough to argue against. Cancer vaccines failed. Patients noticed tumors in the mirror before the immune system seemed to.

Even those who believed, intellectually, that the immune system recognized and killed most mutations of self cells, before those mutated cells ever had a chance to become something we'd call cancer, also conceded that "there is little ground for optimism about cancer"[10] and "the greatest trouble with the idea of immunosurveillance is that it cannot be shown to exist in experimental animals."[11]

There was no data or proof otherwise.

But there were stories.

Through the ages, historians and physicians marveled at these "spontaneous remissions" of cancer,[12] such as the miraculous cure of the thirteenth-century Christian saint Peregrin,[13] later canonized as the patron saint of the disease. These stories or observations seemed like miracles or magic, but to a handful of scientists lucky enough to witness them firsthand, these sudden complete cancer cures were seductive and begged for scientific explanation.

In 1891, William Coley had Fred Stein.

In 1968, Dr. Steven Rosenberg had James D'Angelo.[14]

~

The first hope of therapeutic success comes with the observation of the efficiency of unaided Nature to accomplish cure... These cases, rare though they be, are the sun of our hope.

—Alfred Pearce Gould, "The Bradshaw Lecture on Cancer," 1910

One summer day in 1968, a sixty-three-year-old Korean War vet walked into the West Roxbury, Massachusetts, VA hospital emergency room complaining of severe belly pain. Dr. Steven Rosenberg was the twenty-eight-year-old surgical resident charged with handling whatever came through the door. At first, James D'Angelo presented as just another stubbled vet needing a routine gall bladder operation, but during his medical examination Rosenberg discovered that his patient had a massive scar across his abdomen and an inexplicable medical history.

Twelve years earlier James D'Angelo had been at same hospital with stomach cancer. His surgeons cut out a tumor the size of an orange, only to find smaller nodules like buckshot throughout his liver and abdomen—a death sentence in 1957, as it was in 1968. D'Angelo's grim prognosis had been made worse by a raging post-op bacterial infection. Finally D'Angelo was sent home with 60 percent of his stomach gone—a four-bottles-a-week, two-packs-a-day, stage 4 cancer patient with no expectation of surviving the year.[15] And yet here he was on Rosenberg's examination table twelve years later, very much alive.

Rosenberg asked the VA pathologist to pull D'Angelo's old biopsy slides from storage. The diagnosis had been correct—D'Angelo *had* presented with stomach cancer, an especially aggressive and deadly variety.

Was the cancer still in there, growing slowly, in nonvital organs? Since D'Angelo needed his gall bladder removed, the young surgeon could look for himself. He found nothing in the abdominal wall, and felt nothing in the soft, yielding mass of D'Angelo's liver. "A tumor is easy to identify by touch; it is tough, dense, unyielding, unlike the texture of normal tissues. It seems alien even," he would write later.[16] Twelve years before, according to the detailed surgical notes, the liver had contained several large, dense tumors. Now there were none, and none hiding in any of the other organs either. Rosenberg repeated the examination from scratch. But the cancer was gone.

"This man had a virulent and untreatable cancer that should have killed him quickly," he wrote. "He had received no treatment whatsoever for his disease from us or anyone else. And he had been cured."[17] D'Angelo had beaten his own cancer. There was only one possibility. It had to have been his immune system that had done it.

Which, Rosenberg noted, was exactly what the immune system was *supposed* to do.[18] Immune cells distinguish cells that belong in the body (self cells) from cells that don't belong (foreign, or non-self, cells.) If the immune system overreacts, that's an allergy. If it misidentifies normal self cells and attacks them, that's autoimmune disease. And that was bad. Cancer was supposedly too similar to a normal self cell to be recognized by the immune system; Rosenberg covered that in his years getting an MD and a PhD. But something about D'Angelo suggested otherwise. He didn't have an autoimmune disease, but his immune system had somehow noticed the cancer and had beaten it. There was no other explanation.

This was Dr. Rosenberg's Coley moment, and it would lead to a lifelong obsession. Something that wasn't a miracle had cured this man's cancer.

"Assuming his immune system had destroyed his cancer," Rosenberg wrote, "could the immune system of other people be made to do the same?" D'Angelo's bloodstream seemed to carry the mysterious material of immunity, "not only the white blood cells, but many of the substances that combine to mount an immune response." Was it possible, Rosenberg began to wonder, to transfer those immune response elements to another patient?

~

What Rosenberg did next would be unthinkable today, but both patients involved were agreeable, and Rosenberg was singularly, surgically focused on results, and as quickly as possible. He searched the hospital records and found another patient with stomach cancer and of the same blood type as D'Angelo. When he explained his plan to D'Angelo, Rosenberg remembers, he laughed. "He had gone through a lot worse without helping anybody. He'd be glad to try and he hoped to hell it worked." The patient with terminal stomach cancer hoped more than that. The thin, wheezing skeleton in the bathrobe had once been a gambling man. "He smiled wryly and joked that he had spent his life playing long shots and they had never come in for him yet, and he figured he was due," Rosenberg remembered. If another man's blood might cure him, he was willing to roll the dice.

It didn't work; the transfused blood did no magic, and the patient soon succumbed to his cancer. Rosenberg's experiment had been a failure. Still, he did not doubt what he had seen.

"Something began to burn in me," he wrote, "something that has never gone out."

July 1, 1974, the day after he finished his surgical residency, Rosenberg became the chief of surgery at the National Institute of Health in Bethesda, Maryland, with a staff of nearly one hundred and a lab he would now dedicate to replicating the immune-based cancer cure he'd witnessed in 1968.[19]

Rosenberg wasn't the only researcher focused on nailing down an immunologically based treatment for cancer. But few pushed as hard or accomplished as much as Rosenberg did during those years, and, significantly, nobody else had the nearly blank-check funding from Congress, which helped draw some of the greatest scientific talent from around the globe. For the coming decades, NIH labs at the National Cancer Institute would help keep the field of cancer immunotherapy alive and moving forward. What kept its chief surgeon alive and moving forward seemed to be a healthy ego, burned coffee, and a single-minded focus on curing cancer. At thirty-four years old, this ambitious Bronx-born child of Polish Holocaust survivors was impatient to make his name and change the world. He was going to beat cancer, it was a seven-day-a-week thing, there was no other way. And he was certain that it all hinged on helping immune cells to recognize tumor antigens.

At the time, the scientific consensus was that this was a misguided and futile pursuit, but Rosenberg was one of those who believed the mechanism was already there in the body, waiting to be awakened. As a physician, he had seen patients with compromised immune systems develop cancer at greater rates than those with normal immune systems. As a transplant surgeon, he had seen cancer—probably only a few cells riding invisibly along on a donated kidney—bloom suddenly in the immunosuppressed organ recipient, only to be quashed again when the immune system was restored. He had seen the horrors of graft-versus-host disease, when a patient's immune system rejected a transplanted organ because it seemed foreign. It was a terrible thing, but it showed the power of the immune system. That power, against cancer, would be wonderful.

Other labs around the world were also trying to ignite that wonderful response. Several[20] were pursuing a Coley-like[21] approach to immunotherapy. One involved injecting tumors with a tuberculosis-related bacterium called BCG[22] in hopes that the toxins would spark a broad immune response to the foreign bacterial proteins that might flare into an attack on the tumor itself. It had had some success.

That approach didn't much appeal to Rosenberg. He considered toxins and BCG to be a "broad" and "blunt" approach, an immune "Hail Mary" with "little real intellectual rationale." His idea was to focus specifically on targeting tumor antigens, through a mechanism based on the latest scientific understanding of T cell lymphocytes.

When Rosenberg started in medicine the immunology textbooks didn't even have the word *lymphocyte*. Now they understood that there were two types, B cells, which made the antibodies, and T cells. T's were the immune cells that recognized the unfamiliar proteins on the cells of donor organs, leading to organ rejection and graft-versus-host disease. If the T cell could distinguish one human from another, surely they could distinguish a healthy self cell from its cancerous mutant cousin.

Some mouse studies had suggested that T cells might be able to recognize antigens on cancer cells; Rosenberg chose to believe them.[23] He also believed studies that showed those T cells could be transferred to another mouse surgically implanted with the exact same tumor, killing cancer in one as it did in the other.

Six years earlier, Rosenberg had tried to repeat that experiment at the Roxbury VA hospital using human beings instead of mice. It had failed, terribly. But he still believed in the principle.

Rosenberg believed that D'Angelo had T cells that recognized the antigens of his stomach cancer, much like an immune system that had been inoculated by some cancer vaccine. They apparently didn't do the same job when transfused to another patient, but then those two patients did not have the exact same tumor, with the exact same antigen fingerprints. But what if he could grow T cells specific to a patient's tumor?

At the National Institutes of Health National Cancer Institute, he and his colleagues now attempted to do exactly that, using pigs.[24] It was laborious work, Rosenberg would recall—the procedure required "hoisting them up onto an operating table, anesthetizing and intubating them, scrubbing down exactly as we would for any operation under antiseptic conditions." The surgeons would then place small slivers of tumor samples taken from a human patient into the intestinal lining of these pigs.

After several weeks, the pigs had developed an immune response against the foreign human cancer cell antigens, and built up a clone army of T cells, billions of cells, all specific to recognizing those tumor antigens and killing those tumors. Then Rosenberg's team would harvest the pig's spleen and the lymph nodes closest to the implanted tumor, where the T cell army was concentrated, take them back to the lab, and extract the lymphocytes in strainers. The first test patient was a twenty-four-year-old woman from Pennsylvania[25] with an aggressive cancer and no better options. Even the amputation of her leg had not stopped the spread of her disease.

On November 15, 1977, with approval from the NCI clinical research committee, Rosenberg's team injected 5 cc's of T cells they had previously generated specifically against a sliver of one of her tumors implanted in one of their pigs. She tolerated that test dose, so they proceeded to give her more, ultimately infusing the woman with some 5 billion cells in an hour. This time she developed a high fever and hives, but soon stabilized. The team was hopeful the reaction meant an immune response against the cancer would result, but when she returned several weeks later, her CAT scan showed that the cancer was growing unchecked. The treatment had done no good. It was a crushing failure two years in the making.

~

While one NCI lab had been busy with pigs, three other research scientists[26] also at the National Cancer Institute[27] published a paper outlining an experiment with an unexpected outcome. The researchers had been studying cancer of human blood and bone marrow—leukemia. They'd tried to grow cultures of the disease in the lab, but when they checked the vats, they discovered that they'd accidentally grown a large batch of healthy human T cells instead.

Follow-up investigation suggested that the happy accident had been triggered by a chemical messenger, or cytokine, made by immune cells. The cytokine seemed to act as growth serum for T cells, so they called it "T-Cell Growth Factor"; eventually it would become famous as interleukin-2, or IL-2.[28] For a T cell–focused investigator, IL-2 seemed to be exactly the fertilizer he needed.

If tumor cells did have antigens a human T cell could recognize, they should be able to target and kill it, like any other sick or non-self cell. Something was preventing that from happening. Rosenberg's lab didn't know what that something was—nobody did—but they wondered if maybe they could overwhelm that resistance with a tsunami of T cells.

We all have about 300 billion T cells circulating through our bodies, each of them a lottery ticket randomly tuned to every possible antigen-recognizing combo. While that might sound like an enormous number, keep in mind that only those T cells that recognize the antigen fingerprint of an infected or a sick cell activate. And there's no way for the immune system to predict what that antigen fingerprint might be. As a result, those 300 billion combinations need to account for—and potentially match up with—every possible antigen that nature might throw at us. That means that in this antigen lottery, of those 300 billion possible combinations, at most only a few dozen T cells have the same winning ticket—the exact right receptor capable of recognizing any one antigen, should it happen to show up.

But what if you boosted the odds by boosting the number of T cells? Surely one of the 300 billion T cells had the winning receptor that happened to match the tumor antigens. Ideally, a researcher would figure out which one matched, make a billion copies with the interleukin-2 fertilizer, and infuse them back into the patient. At the very least, if she could induce *all* of those 300 billion T cells to replicate, she'd end up with even more versions of all the possible combinations—including more copies of the one that happened to match the tumor antigen. Instead of twelve winning tickets, she'd have twelve million.

Rosenberg met with the authors of the IL-2 research. Then, on September 26, 1977, he tried it in his own lab, following the borrowed recipe for making IL-2 from mice. His lab added the powerful potion to a culture of ten thousand T cells. When they checked five days later, the mass had swelled to 1.2 million cells.

More was good, but were they still killers? And were any of them killers that could recognize and kill cancer? And would they be killers not just in a test tube, but in a living animal—a barrier that had stumped many hopeful immunotherapies over the years? And finally, the ultimate barrier: Would all that translate to humans?

Those questions would consume the next years of the many talented young scientists who passed through these government-funded laboratories. The work was slowed considerably by the difficulty of getting sufficient quantities of IL-2, a time-consuming process that was far harder on the mice than the researchers. By the early 1980s that dynamic changed, with the advent of new technology in genetic engineering and molecular biology. For the first time, researchers could manipulate the DNA blueprints of bacteria, inserting genes that turned them into living chemical factories. A number of biotech companies jumped into the race to use recombinant DNA to produce wonder drugs. IL-2 was an afterthought; at the time, the goal was to mass produce a cytokine called interferon.

~

Like most science stories, the story of interferon starts with a mysterious observation: monkeys infected by virus A (in this case, the Rift Valley fever virus) were afterward resistant to infection by virus B (in this case, the yellow fever virus).

The concept of inoculation and vaccines had long been familiar, but what was observed in these monkeys in 1937 was something new. This wasn't inoculation, as the two viruses did not appear to be related to each other. Some different sort of biological mechanism seemed to be at work. Follow-up experiments showed that the mysterious phenomenon extended beyond these monkeys or these viruses specifically. In various cells, and all manner of animals, exposure to one virus (usually a weak, nonfatal sort) somehow interfered with the ability of a second, potentially fatal virus to infect the host cell.

Viruses are essentially just genetic material in a tiny crystal syringe. They cannot reproduce on their own; instead, they inject their genetic payload into a host cell. The virus's genetic blueprints reprogram that cell's genetic machinery to stop making proteins that help the host and to start producing virus parts instead. Somehow, the experiment suggested, prior exposure to a virus interfered with that master plan, the way a large radio tower crowds out smaller stations on the dial. They called the phenomenon "interference."

Throughout the 1940s and 1950s, the search for the essence of viral interference was the most intriguing quest in biology, drawing a generation of young scientists to its study. If this "interferon" existed as a hormone-like liquid, it was hoped, it might hold the power to vanquish disease.

That hormone-like liquid was finally described in 1957 by researchers Alick Isaacs and Jean Lindenmann, as found in the membranes of chicken cells they'd cleverly infected with a flu virus.[29] The resulting clear, powerful syrup turned out to be a previously unseen type of protein—one of three major classes of cytokines produced by animal cells in response to viral attack and, in some cases, the presence of a tumor.

Interferons (IFNs) were the first cytokine to be heralded—some would say hyped—as a potential magic bullet in the war against disease, including cancer, and they wouldn't be the last.[30] The first eyedropper-sized batches were painstakingly squeezed from white blood cells centrifuged from donations collected en masse by the Finnish Blood Bank and mashed through incrementally finer porcelain filters. The process was messy, but the result was, for a time, the most precious commodity on earth.

That changed with the invention of recombinant DNA technology. By 1980 scientists could manipulate the DNA blueprints of yeast cells well enough to begin pumping out interferon proteins like a brewery. Finally, researchers had sufficient supply to begin testing the reality of interferon against nearly four decades of hype, and hopes were perilously high for what *Time* magazine promised on its March 31, 1980, cover story was the "penicillin of Cancer."

IFN would never quite live up to the pent-up enthusiasm. The research presented good science and provided important new bio-chemical insights[31] and even a few practical medical applications. But ultimately what would be publicly remembered was how far it missed the "penicillin" mark as a magic bullet cure for the C word. In the end, hope for IFN had spiked and crashed in the course of a *Time* magazine news cycle, one more disappointingly premature cry of eureka in the troubled history of immunotherapy.

But in 1980, that disappointment was hardly imagined. The excitement over interferon had fueled a speculative boom in the hand-ful of biotechs that could engineer and produce the valuable stuff, companies that soon searched for other scarce biochemicals to mass-produce. And at the time, nothing was more scarce, or potentially more important or lucrative, than the one the brilliant young post-docs in Steve Rosenberg's lab were clamoring for: intereuken-2 (IL-2).

~

IL-2 is an incredibly powerful cytokine, effective even when diluted at a ratio of 1:400,000 (one part IL-2 to 400,000 parts inert solution). IL-2 also degrades quickly, preventing powerful and specific immune battle commands from echoing dangerously around the body after they're no longer relevant. Its half-life of less than three minutes[32] wasn't nearly long enough to accomplish the job Rosenberg and his colleagues had in mind. To keep conduct-ing experiments providing the immune cells with the growth sig-nal, especially during the critical period following recognition of tumor antigen and activation, even more IL-2 would be required. That meant more lab hours, and many, many more mice. Finally, on June 12, 1983, the head researcher at a Stanford biotech spinoff called Cetus surprised Rosenberg as he was about to board a plane from a conference, handing him a test tube full of recombinant gene-made IL-2. Rosenberg secured the vial of the most precious stuff on earth in his jacket pocket. "I tried to hide my excitement," he remembered. It's difficult to imagine he was convincing as he cautiously boarded a plane with an amount of IL-2 that dwarfed all previous supply.

The vial survived the trip, facilitating experimental ventures into levels of T cell growth previously considered impossible; what was more, Rosenberg was promised, even greater amounts would soon be available. As production ramped up, those test tubes would become flasks, then buckets; researcher Paul Spiess later calculated that the unused drop of recombinant IL-2 wasted at the bottom of a test tube represented the amount of natural IL-2 that would have formerly required the sacrifice of 900 million mice.

"I had felt as if there was a powerful machine at my disposal, that its engine was ready to roar, but that I could not find the key to it," Rosenberg later recalled. "I had wondered if IL-2 was that key. Now I would find out."

As promised, his lab had taken a methodical approach to the experiments, all based on the yet-unproven premise that T cells could recognize the antigens of cancer cells in humans. They now had two main approaches to using the IL-2 to grow a T cell army that might overwhelm cancer. One approach was to remove a patient's T cells, fertilize them with IL-2, then reinject that forti-fied T-cell army into the patient. Another approach was to feed IL-2 directly into the patient's bloodstream to fuel and support any response their immune systems might naturally initiate.

At sufficient doses, both approaches worked in mice. But by November 1984 it was clear that, once again, what worked in mouse models didn't translate to people.[33]

"Perhaps for the first time, at least part of me began to doubt the path that I had begun to follow," Rosenberg would later allow. This was a rare confession of self-doubt from the hard-charging surgical chief, as well as a massive understatement of the human stakes and scope of his failure. Congress wanted results from the hundreds of millions it had spent on the war on cancer; Rosenberg was at a government lab, spending public money on pigs and mice and racking up a record of sixty-six consecutive "failures"—sixty-six human beings he'd gotten to know, tried to help, and failed to save trying one experimental approach then the other.

Finally, on November 29, 1984, desperate to make something work, he tried both approaches at once, and at double the previous dosage of this powerful cytokine.

His team injected a bolus of IL-2 fertilized T cells back into a woman named Linda Taylor, a former Navy brat and military attaché who'd suffered a relentless melanoma that was unresponsive to other treatment. It took nearly an hour to drip the mass of 3.4 billion cells into her arm. Then she was given large injections of IL-2 to sustain the immune action, over 40 million units daily, for six days.

Taylor responded to the combination treatment. Within a few weeks, the tumors began to get smaller and squishy—under the microscope they revealed necrotic tissue, dead tumor. By March of the following year, Taylor's scans showed no cancer at all. "It had disappeared," Rosenberg reported. It was working. He felt a new urgency to continue this combination technique, to "push harder," with more patients.

The results of that larger study were mixed. The treatment still did not help most patients, and the side effects could range from debilitating to deadly. Rosenberg described how, for him and the staff, a visit with a responding patient was a special thrill quickly tempered by the patient in the next bed who was not responding to the treatment at all and was only closer to death from the side effects. They had no idea why a treatment that worked in some patients failed entirely in others. And so while the treatment provided data, and helped some patients, it didn't definitively prove anything. The IL-2 treatment seemed to clear cancer in some patients. It definitely proved fatal to others. This outcome was emotionally and physically exhausting. Even some of those who survived both treatment and cancer suffered traumatic flashbacks for years afterward.

But Rosenberg maintained that the numbers were not uncommon for cancer trials. The patients in these experiments knew that while there were definitely risks inherent in the testing of an experimental medicine, the mortality rate for doing nothing was 100 percent. Still, some at the NIH wanted to cease the treatments. Rosenberg vowed he wouldn't stop "until they make me." Eventually, they did exactly that.

This was a dark time for Rosenberg, but he believed they needed to continue to test the possibilities of the therapy and announce the findings, good and bad. Further to that, the president of the National Cancer Institute, the pioneering chemotherapist Dr. Vincent T. DeVita,[34] was under pressure from Congress to justify the millions spent on the war on cancer with some, or any, proof of success. That fall, the *New England Journal of Medicine* accepted a paper from Rosenberg et al. that cautiously reported the result from twenty-three patients. That paper was scheduled to be published in December 1985, but sent, under embargo, to health reporters a week early, so they might prepare. That, Rosenberg would later write, was a mistake.

Rosenberg's scientific paper was beaten to the news stands by a feature story in *Fortune* magazine. The cover showed a photo of a test tube of medical-looking liquid labeled "Cetus Corps tumor-zapping interleukin-2."

The cover line read: CANCER BREAKTHROUGH.

Rosenberg says that his reaction was apoplexy. "Cancer breakthrough," he declared, was exactly the sort of hyperbole serious scientists wanted to avoid; the *Fortune* cover was irresponsible and misleading. Yes, a small minority of patients responded completely to the treatment, but they couldn't predict who those patients would be, or why it worked in some patients and on some cancers, or why it failed in others. And some of the responders had relapsed fatally. "We had not cured cancer," Rosenberg declared. "We had only detected a crack in its stone face."

Nevertheless, between the *Fortune* cover and the *NEJM* "special issue" a week later, the breakthrough genie was out of the bottle. All the major networks ran the breakthrough story on the evening news. The next day it was on the front pages of the *New York Times*, the *Los Angeles Times*, *USA Today*, the *Washington Post*, the *Chicago Tribune*, and hundreds more papers around the world. Rosenberg agreed to a walk-through of the wards with Tom Brokaw, hoping to course correct the sensationalized *Fortune* cover, but that story had already set the "breakthrough" tone. The newsweeklies followed, with major coverage in *Time* magazine and with Dr. Steve Rosenberg smiling benevolently from the cover of *Newsweek*.

Now the NIH was being bombarded with interview requests from journalists, and hundreds of calls a day from cancer patients across the world. Switchboards at cancer centers across the country were soon being flooded by hopeful, desperate patients. Looking back at the hype, Rosenberg was confounded. He'd published the results of his work but never declared he'd made a breakthrough. Perhaps the media frenzy was because he was already a face on the nightly news: not only as chief of surgery at the NIH but also as the surgeon who had operated on President Ronald Reagan, and then bluntly told the nation on live TV what none of his press secretaries dared say: "The president has cancer." That press conference and the backlash to his blunt honesty had surprised him. This was far worse.

"With an increasing sense of urgency, I tried to play down the expectations," Rosenberg would write later. But Rosenberg lived for his work, and several of his colleagues felt that even as he tamped down the flames, he seemed to appreciate at least some of their heat and light; certainly, they illuminated the focus of his life's work, and brought attention to it. In his interview with *People* magazine, in which he would feature as one of their "people of the year," Rosenberg referred to his lab's findings as "the biggest advance in cancer for 30 years." Even as he rebuffed the breakthrough angle on his immunotherapy, he also sometimes referred to it as such, using the B word.

One Sunday morning, Rosenberg and DeVita set aside a few hours to appear on CBS's *Face the Nation*. In conversation with the staff before the show taped, DeVita mentioned one of the patient deaths, a particularly difficult and personal episode that underscored the need to temper the sensational headlines. That death hadn't been mentioned among any of the twenty-three patients Rosenberg had reported on for the *NEJM*, and it hadn't been part of any previous news report. It was, in short, a scoop, and a few minutes later Lesley Stahl, the show's host, popped in to say hello, asking in an offhand manner, was it true there had been an IL-2–related death?

Rosenberg had never spoken of the man, a patient named Gary Fowlke, publicly. He found the notion of "offering the press a running scorecard on patients" offensive, and didn't believe the press understood how dangerous cancer treatments—any cancer treatments, and especially experimental ones—truly were. Daytime TV certainly wasn't the place to publish scientific information. And yet, despite all that, it was also true—in all the coverage, he hadn't mentioned that death or the horrifying side effects.[35]

Rosenberg says he decided to beat Stahl to the punch and bring up Mr. Fowlke's death in the clinical trial before Stahl had a chance to ask about it. But the damage had been done. The sensational headlines around Rosenberg's experimental results had given most of the world their first exposure to cancer immunotherapy. And as high as public hope had soared on that exposure, it now suddenly came crashing back to earth with a vengeance.

"There may be a balance scientists can reach in publicly discussing a scientific development—a balance between the public's right to know and scientists' fears that the public's lack of expertise will lead to misunderstanding or unrealistic expectations," Rosenberg would reflect later. "But in that case I failed to reach it."

But the peaks and valleys of sensational coverage didn't change the data, or the results Rosenberg's lab had elicited from his cancer patients. And so, despite the uncertainty over the exact biological mechanism, on January 16, 1992, the FDA approved IL-2 for patients with advanced kidney cancer. It wasn't a cure, or even a frontline, first-choice approach. But it was, Rosenberg proudly noted, the first approval in the United States of a treatment for cancer that acted solely through stimulating the patient's immune system.[36] Many researchers now believe that when combined with the newest cancer immunological advances such as checkpoint inhibitors, IL-2 may prove to be more important than even Rosenberg had then realized. But perhaps most important was the glimmer of proof the NCI labs had provided the world. Cancer immuno-therapy *could* work, and in fact had. The underlying science was still poorly understood. Rosenberg's methods and success rates proved very difficult to reproduce,[37] and a great deal of basic immunological research was yet to be undertaken. But there it was, in black-and-white data, and in living patients as well. Rosenberg paraphrases Winston Churchill when assessing the impact of these IL-2 stud-ies; it was neither the end nor the beginning, but rather, the end of the beginning of the cancer immunotherapy story.

Those glimmers inspired a handful of talented young research-ers to enter the field, and they sustained the handful that remained. For decades to come, the army of scientific talent that passed (and still passes) through the NIH laboratories would read like a who's who of those leading advances in the cancer immunology field.

But for everyone else—the oncologists trained when *Coley* was a dirty word, the researchers who had been suspicious of the unre-producible results, and most especially the general public for whom Rosenberg had been the face, and interleukin-2 the promise, of sal-vation from an uncurable disease—it was a disaster. Cancer immu-nology became the science that cried "breakthrough" on the cover of *Time* once too often. Immunotherapy's moment came and went, and the spotlight with it.

It was the 1990s, and DNA manipulation was the apparent future for potential cancer cures. Oncogenes, genes which, when mutated, increase the likelihood of a cell becoming cancer, had been identified, as had suppressor genes, which seemed to work against those destabilizing mutations, and researchers sought to target them. Soon these efforts were joined by targeted therapies and "inhibition pathways,"[38] small molecules that targeted the metabolic means by which cancer created its blood supply or requisitioned the fuel it needed to grow and divide. These were cancer therapies that, like radiation, chemotherapy, and surgery, directly targeted the disease instead of acting upon the immune system. That made sense to people, and they worked, to a degree. The new scientific technology made those drugs easier and cheaper to make and more successful than before, adding weeks or months to cancer patients' lives. They also made headlines, eclipsing immunotherapy research and outcompeting it for R&D funding. After the breakthrough, "bust" became the next immunotherapy story line.

"We search where there is light," said Goethe. And the promise in cancer immunotherapy was still just an occasional flash in the darkness. For the best and brightest young scientists, cancer immunotherapy was explicitly a no-go zone as a career choice. Most of the generation graduating in the late '80s and through the '90s gravitated toward better-funded and more hopeful fields of scientific inquiry. Some went into developing new classes of chemotherapy, or radiation oncology. Many went into "pathway inhibition" science. And cancer doctors maintained the traditional cut, burn, and poison treatments they had been taught by the generation before them, the only weapons they could truly trust.

Basic but essential immunotherapy research was left to the handful of true believers, researchers still quietly chipping away, like Lloyd Old, Ralph Steinman, and others. Steve Rosenberg, meanwhile, had moved on from IL-2 to new targets and technologies, following Dr. Phil Greenberg's lead in figuring new and better ways to grow up and transplant armies of T cells that could recognize and kill tumors.[39] Though you wouldn't have guessed it by looking at the nearly empty cancer immunotherapy presentations at the national cancer conferences, populated by the same faces, often from underfunded labs, year after year, there were still more avenues left to try for making a successful cancer immunotherapy. What most of those avenues had in common was the immune cells cancer immunologists still believed could recognize tumor antigens and kill cancer: T cells.[40]

But that raised the now familiar question: If T cells could recognize cancer antigens (they can), and if Greenberg and others had been able to grow and stimulate a T cell army that recognized tumor antigens and would attack cancer (they did), why didn't cancer patients experience an immune response to cancer *without* such intervention? If the immune system could see and kill tumors, why didn't it? Why did we have cancer at all?

There were two possible answers: either the immunotherapists were wrong, or something was still missing from the equation.

The questions were interesting. Dr. Rosenberg was more interested in pushing experiments theory into the clinic as quickly as possible, even if that meant outpacing the basic immunological research that might make sense of the results. But it was clear that something was missing, something undiscovered, like an unknown puzzle piece, that prevented T cells from getting "activated" against cancer, or shut them down before they could complete the job. This wasn't an observation about the immune system or disease in general; that mysterious something seemed to happen *only* when the immune system interacted with cancer cells.

If you were a chemotherapy-trained oncologist or a molecular biologist, the idea of an elusive *something* sounded quaint, and not very scientific.[41] Which meant cancer immunotherapy wasn't a legitimate science. You believed, or you didn't; it all came down to which studies you chose to believe, and how you decided to interpret them.

Immunotherapy cynics (constituting the vast majority of the people working with cancer, the immune system, or both) believed that the elusive "something" that kept cancer immunotherapy from working was called "reality": Cancer and the immune system didn't interact, they had nothing to say to each other, and the conversation could not be forced. Any anticancer effects of interferon or IL-2 or BCG were surely just the T cell recognizing an antigen from a virus that had infected a cell, and led to cancer. Nobody was arguing that T cells didn't recognize virus-infected cells; they did. And some cancers were known to be more likely after infection by a virus (such as HPV).[42] Here was a model that fit the fact pattern, shaved by Occam's razor; Rosenberg, they maintained, had simply misinterpreted what he'd seen. The antigens of a cancer cell just weren't non-self enough to be recognized as foreign by a T cell. If they were, one could make a successful cancer vaccine. And yet, no such vaccine then existed.

Cancer immunologists could argue about this all they wanted, they could point to the glimmers. But at the end of the day, they didn't have the biology to back up their arguments. Really, there was only one successful counterargument that could be made: discover that certain *something* that explained the problem with cancer immunotherapy and allowed T cells to reliably recognize, target, and kill cancer cells. And in the race to do that, the most successful would be the ones who weren't even trying.

Chapter Four

Eureka, Texas

Chance favors the prepared mind.

—Louis Pasteur

The one who finally found the *something* was a hard living, harmonica-playing Texan who wasn't even really researching cancer.

Jim Allison looks like something between Jerry Garcia and Ben Franklin, and he's a bit of both: a musician and scientist who sugars his impatience and raw intelligence with beer twang and humor. More than anything, he is a curious and careful observer who doesn't seem to give a damn about much else—a basic science researcher happy to be wrong ninety-nine times to be right once.

Allison outgrew his hometown of Alice, Texas,[1] in high school, after he was forced to turn to correspondence classes for an advanced biology class that dared mention Charles Darwin. That class was in Austin, home of Texas's best public university and its most happening music scene. The combo suited Jim Allison perfectly, and after high school he moved there for good, seventeen years old and bound to be a country doctor, as his dad was.

The stretch between 1965 and 1973 was a good time to be young and musical in Austin.[2] Jim played the blues harp, well enough that he was in demand. He'd play honky-tonks in town, or play for Lone Stars in Luckenback, where the new breed of outlaw country players like Willie Nelson and Waylon Jennings roamed the earth.[3] That was fun; premed meanwhile felt like memorization for nothing.

In 1965 he switched to biochemistry, and traded memorization for a biochem lab, working with enzymes for his PhD. The enzymes he was studying happened to break down a chemical that fueled a type of mouse leukemia.[4] As a biochemistry PhD candidate, Allison was supposed to figure out the biochemistry of how those enzymes worked.[5] But he was also curious about what had happened to the tumors.

"So I was reading all this immunology stuff in the library," Allison says.[6] In the experiment, the enzyme eventually robbed the tumor of all its fuel, and the tumor went necrotic and "disappeared"—just another dead cell mass to be cleaned out by the macrophages and dendrites. But from his reading Allison knew those blobby aomeba-like cells weren't all garbagemen; they had recently been discovered to also be frontline reporters, carrying updates on the constant battle against disease. Those updates were contained in the dead and diseased cells they gobbled into short protein fragments—the distinctive antigens of the disease pieces. Macrophages (and dendrites) were first on the scene, everywhere, embedded. When they found something interesting they brought pieces of the non-self proteins they'd gobbled back to the lymph nodes to show them around. (Lymph nodes are like Rick's in Casablanca. Good guys, bad guys, reporters and soldiers, macrophages, dendrites, T and B cells, and even diseased cells, everyone goes to Rick's.)[7] That's how B and T cells found their antigen, and activated.

What the macrophages were doing with the dead tumor tissue in his mice gave Allison a thought: That's sorta how a vaccine works, right? A vaccine introduced the immune system to a dead (inoculated) form of a disease so that the immune system could prepare a response against that disease—build up a clone army of T cells specific to it, so that even an invading force of that disease would be evenly met. And wasn't that what he'd done, by killing a tumor that macrophages cleaned out? Weren't dead tumor cells, gobbled and presented by macrophages, something like a vaccine? So, he wondered—did that mean his experiment had, in a round-about way, vaccinated his mice against this specific form of blood cancer? Were they now "immune" to this cancer?

"Just for the hell of it, I was setting up another experiment, and I decided that since I had these mice that were cured—who were just sitting there, eating—I would inject them with the tumor again, but not treat them with the enzyme this time, and see what happened." This wasn't the experiment—he hadn't asked permission, he didn't write a protocol, nothing. He simply shot from the hip. And what happened was—nothing. "They didn't get tumors," Allison says. "I went back and injected them with ten times as much and they still didn't get tumors. I injected them with another five times more, and they still didn't get tumors! Something was happening here, something amazing!"

As a casual one-off, the experiment hadn't proved anything ("People talked about doing it in humans, you know, just taking your own tumor and mashing it up somehow and injecting it back, but it doesn't really work that easily"), but it had provided Allison his first glimpse of the mystery and potential of the immune system. It was the most interesting thing he had seen. Now, he wanted to study that, first in a postdoctoral position at the Scripps Institute in San Diego,[8] and then at a little lab MD Anderson Cancer Center was opening near the town of Smithville, Texas, "an economic stimulus thing from the governor," Allison says, on donated land and with state money.

"It was pretty weird," Allison says. "It was in the middle of an eighteen-acre state park,[9] and they'd just set up some lab buildings and hired six faculty members to go out there. We were supposed to study carcinogenesis [how cancer starts]. I didn't know anything about that." But he had picked up some immunological techniques that helped those experiments work. Meanwhile, Allison says, MD Anderson sort of forgot about them.[10] "So they pretty much left us alone." This was Allison's kind of place—for now, anyway. His colleagues were bright, enthusiastic scientists his own age— the oldest were in their thirties—who kept beer in the lab, worked late, helped each other with their experiments, and pooled intellectual resources.

The setup was sweetened by a total lack of teaching or administrative responsibilities, a Norton Commando 850 motorcycle, and enough NIH and NCI grants to pursue what Allison was really interested in—the study of a recently recognized lymphocyte, the T cell.

"It was a fantastic time in science because immunology had just been this poorly understood field," he says. "I mean, everybody knew we had an immune system, because there were vaccines. But nobody knew much about the details of anything."

One of the things nobody knew was how a T cell recognized a sick cell in the first place. Allison read every academic paper he could on the topic, then read the papers cited in them. "At first, I'd think, I'm an idiot, I can't understand this. Then, I thought, No, *they're* idiots—they don't understand what they're talking about!"

There were plenty of theories about how a T cell recognized antigens.[11] One prevailing theory was that each T cell had a unique type of receptor (a specific arrangement of proteins that extended from the T cell surface) that "saw" a specific antigen expressed by a sick cell, homing in and fitting something like a key into a lock.

That was a reasonable theory, but nobody had actually found one of the receptors. If they existed, there should be a lot of them, scattered among all the yet uncounted proteins that stuck from the T cell surface (there are so many that new ones are given numbers, like newly identified stars).[12] Those "receptor" proteins would be molecules built in some sort of double chain-like configuration. Several labs were quite convinced that it would look just like it did on B cells. Which, Allison thought, was stupid.

"People from Harvard and Johns Hopkins, and Yale, and from Stanford were already claiming they had a molecule that was the T cell receptor," Allison remembers. "Most of them, because B cells make antibodies, figured that in T cells the receptor had to be an antibody-like thing, too."[13]

Whatever it looked like, if you could find it, in theory you could manipulate it. Control the T cell receptor and you might control what the immune system's killing machine targeted. The result could have massive implications for humanity, and make a massive name for whoever found it.

Allison believed that T cells weren't just a version of B cells, not just killer-B's. If T cells existed (they did) and were different from B cells (they were), then those differences were the point. The molecular structure of the receptor that allowed T cells to "see" their specific antigen target was one of the key points of differentiation from B cell receptors; it would look different, because it worked differently, and did a different job.

The idea came in a flash while he sat in the back row of a lecture on the topic, listening to a visiting Ivy League academic. Suddenly, it seemed so obvious: if he could find a way to compare B cells and T cells, devise a lab experiment that put one against the other and let their redundant surface proteins cancel each other out, the receptor should be the molecule that *didn't* cancel out. Essentially, he was looking for a needle in a haystack, and his idea was to set fire to the haystack and sift the ashes. Whatever was left would be the needle he was looking for.

He hurried to the lab and got to work. "It was a success, the very first time," he says. "So, now I've got a thing that's on T cells but not on B cells, not on any other cells[14]—so, that's gotta be the T cell receptor!" He showed the receptor was a two-chain structure—an alpha and a beta chain, and he wrote it up in a paper.

Allison was hoping to be published by one of the leading peer-reviewed research journals.[15] But nobody at *Cell* or *Nature* or any of the A-list peer-reviewed journals was willing to publish the findings of this junior academic from Smithfield, Texas. "Finally, I ended up publishing the results in a new journal called the *Journal of Immunology*." It wasn't *Science* or the *New England Journal of Medicine*, but it was in print, and in the world.[16]

"At the end of the paper, I said, 'This might be the cell antigen receptor, and here are the reasons why I think that it *is* the T cell antigen receptor,' and I just listed it out, all the reasons." It was a bold announcement regarding the biggest topic in immunology. "And nobody noticed it," Allison says. "Except in one lab."

That lab was headed by eminent biologist Philippa "Pippa" Marrack at UCLA San Diego. Her lab (shared with her husband, Dr. John Kappler) hadn't identified the T cell receptor yet, but they had a scientific technique that could verify if Allison's results were correct. Dr. Marrack reproduced Allison's experiment and got an exact hit on the protein Allison had identified—and only on that protein. It was a shock, especially coming out of a lab Marrack had never heard of. Allison said she called and told him she was organizing a Gordon Conference—elite, closed-door gatherings something like the Davos of science. She invited him to present at the meeting; Allison had a sense he was being invited into the big leagues.

The Gordon meeting helped put the brash young scientist on the academic map, and won him an appointment as a visiting professor at Stanford University. Now that the T cell antigen receptor (TCR) had been identified, and its two-chain molecular structure had been described, the race was on for the greater prize: the blueprints for those proteins, as encoded in genes in the T cell DNA.

"At that point, people had just figured out how you could work with DNA and clone genes, so now, everybody was trying to clone this [T-Cell receptor protein] gene," Allison says. "It had been the holy grail of immunology for twenty, twenty-five years, and nobody had solved it. Everybody was scrambling, man, it got ugly. I mean, everybody realized there was a Nobel Prize at the end of it."

That August, Stanford immunologist Dr. Mark Davis made an unscheduled speech at the big tri-annual immunology world congress in Japan, announcing that his lab had located the gene for the T cell receptor beta chain in mice. The following year he published the confirming details in the prestigious British journal *Nature*, back-to-back with a paper by renowned Canadian geneticist and biological researcher Dr. Tak Mak, who had successfully identified the T cell receptor beta chain gene in humans. That left the gene for the other half of the T cell receptor, the alpha chain. Davis, along with his collaborator and wife, Dr. Yueh-Hsiu Chien, were in the audience when that achievement was announced during a slide presentation by MIT immunologist Susumu Tonegawa.[17] Davis had shared his lab's gene-cloning technique with Tonegawa a few years earlier; now he felt like he was paying the price. On the plane ride home, Chien told her husband she recognized the slide of the barcode-like "fingerprint" Tonegawa had announced as coding the alpha chain. Davis smelled opportunity. They rushed back to their lab, pushed around-the-clock research into the gene Tonegawa's slide seemed to have identified, and put a written paper on the subject onto a 7 p.m. DHL flight to London, where it was hand-delivered to the editor's desk at *Nature*. Tonegawa's own paper on the alpha chain gene arrived at the same desk days later.

While both articles, with nearly identical titles and announcing the same discovery, were published back-to-back in the November 1984 issue,[18] technically the Davis and Chien paper had hit the desk first, giving them the honor and the citation in biology textbooks forever.[19] Two years later, Susumu Tonegawa would be awarded the 1987 Nobel Prize in Medicine, citing his earlier groundbreaking work on B cell genes. To date, nobody has received a Nobel for the T cell receptor gene. Afterward, Tonegawa left immunology to study the molecular basis of what and how we remember, and what and how we forget.

~

"Anyway, we cloned a lot of stuff," Allison says. "But none of it was right. At the end of it, I was invited to give a seminar at [University of California,] Berkeley. It was kind of controversial because I hadn't been at the big labs. I hadn't been at Harvard. I lacked the pedigree of most faculty at places like Berkeley." Which was why it blew his mind two weeks later when Berkeley offered him a full-time job,[20] covered by a massive grant from the Howard Hughes Medical Institute. Allison would have a lab and postdoc salaries, and he could research whatever he wanted. He didn't need to teach, and the money might last forever with no strings. His only obligation would be to go to the HHMI headquarters every three years and give a twenty-five-minute talk in front of fifty of the top scientists in the world and present his work on T cells.[21]

Allison's work at Berkeley would have the benefit of a far better understanding of T cells than when he had first started fixating on them a decade before. Now it was widely accepted that there there were different kinds of T cells, with different specialties for coordinating an immune response against disease. Some "helped" immune response by sending out chemical instructions, via cytokines, like a quarterback calling plays. Others, the killer T cells, killed infected cells one-on-one—usually by chemically instructing those cells to commit suicide. The processes above, and more, were set in motion only when a T cell was "activated." *Activation* is the beginning of the adaptive immune response to disease; until then, the T cells are just floating around and waiting. So what activated T cells? What made them start mobilizing against disease?

"We thought that the T cell antigen receptor was the ignition switch," Allison says. That was the natural assumption.

It was only after they'd identified the T cell receptor that they realized, nope, that wasn't quite right, either.[22] They could get the T cell receptor to "see" the foreign antigen of a sick cell; they bound like lock and key. But the antigen key wasn't enough to turn on a T cell.[23] It wasn't the "go" signal.

"When I learned that, I said, 'Oh, wow, this is cool, T cells are even more complex,' you know? It just added more to the puzzle. It made it more fun."

If keying the T cell receptor with an antigen wasn't the only signal needed to turn on a T cell, that meant there had to be another molecule, maybe several, required for costimulation.[24] Maybe the T cell required two signals—like the two keys for a safe deposit box, or how, when starting a car, you need to key the ignition and also press the gas pedal to make it go. But where was the T cell's gas pedal?[25] Three short years later, they found it, another molecule on the T cell surface called CD28.[26] *CD* stands for "cluster of differentiation," which is sort of like calling it "a thing that's clearly different from the other similar things around it."

CD28[27] was the second signal to turning on T cells.[28] That was important, but, as Allison and other researchers quickly discovered, it also wasn't that simple. Presenting the right antigen key to the T cell receptor *and* costimulating CD28 did start up the T cell, but when they did that in mice models, often the T cell just stalled out. It was as if they'd found the key to the ignition and the gas pedal, but a *third* signal was still necessary to make the T cell "go." So now they went hunting for that.

One of Allison's postdoctoral students, Matthew "Max" Krummel, compared the structure of the protein CD28 to other molecules, looking for something similar in a sort of computerized book of molecule mug shots—"the gene bank, that's what we called it at the time," Allison says. The idea was that if you found a molecule that looked similar, maybe it did similar things and was related. Krummel soon found another molecule with a close family resemblance to the part of CD28 that stuck out of the cell, the receptor part.[29] The molecule had recently been identified, named, and numbered.[30] It was the fourth cytotoxic (cell-killer) T-immune cell (lymphocyte) identifed in the batch, so Pierre Goldstein, the researcher who'd found it, called it cytotoxic T-lymphocyte-associated protein #4—or CTLA-4 for short.[31]

Meanwhile, researchers Jeffrey Ledbetter and Peter Linsley were working on the same third signal problem at the Bristol-Myers Squibb research campus in Seattle. "Linsley made an antibody to block CTLA-4," Allison recalls. The group published a paper concluding that CTLA-4 was a third "go" signal, another gas pedal on the T cell that had to be activated for immune response.[32] Having another researcher beat them to making the antibody was disappointing, and especially disheartening to Krummel—he'd just spent three years working on the antibody as his intended thesis project—but Allison decided to proceed with more CTLA-4 experiments anyway. There was always more to learn. Besides, he wasn't totally convinced that Linsley et al. had completely solved the T cell activation mystery.

"I knew there were two ways you can get something to go faster," Allison says. "One is to press on the gas pedal. The other is to take off the brake." Allison says Linsley's group had only devised experiments consistent with CTLA-4 being another "go" signal, essentially a second CD-28. "I said, 'Let's do the experiments consistent with [CTLA-4] giving an "off" signal,'" Allison says. "Sure enough, that's what we found out. CTLA-4 was an 'off' signal."[33]

Allison's lab now had a fairly complete picture of the steps required for T cell activation against disease. First, the T cell needed to recognize the sick cell by its unique protein fingerprint; in other words, it needed to be presented with the antigen that matched its T cell receptor (TCR). Usually it was a dendrite or macrophage that did that presenting. Binding to that antigen was like turning the key in an automobile ignition.

The other two signals (CD28 and CTLA-4) were like the gas pedal and the brake on that car. CTLA-4 was the brake—and it was the more powerful of the two. You could press both (and in experiments, Krummel found that was a crude way of controlling the activation rate), but if you floored both, the brake overruled the gas pedal, and the T cell wouldn't go, regardless of everything else. Enough stimulation of CTLA-4, and immune response stalled out.

If all this sounds complicated, it's because it is, on purpose. Allison's lab had discovered an elaborate safety mechanism, an aspect of the larger framework of checks and balances that prevents the immune system from going into overdrive and attacking healthy body cells. Each safety is a sort of fuse that gets tripped if a trigger-happy T cell is programmed to target the wrong antigen, such as one on normal body cells. It was a way of repeatedly asking, *Are you sure about this?* before T cells turned into killing machines.

Proper triggering of immune response against pathogens is what keeps you healthy. However, pedal-to-the-metal immune response against self cells is autoimmune disease: multiple sclerosis, Crohn's disease, some forms of diabetes, rheumatoid arthritis, lupus, and more than a hundred others. They happen often, even with this elaborate feedback system in place. And so the double-check, double-signal mechanism of T cell activation is only one of many redundancies and fail-safe feedback loops built into immune response. Those "checkpoints" on T cell actuation hadn't been guessed at.[34] But now Allison's lab, and, simultaneously, the lab of Jeff Bluestone at the University of Chicago, had found one of those checkpoints.[35] Bluestone was focused on ways of placing this new discovery in the context of organ transplants and diabetes, tamping down unwanted immune response. But Allison had a different idea where he'd like to stick it.

Biology was interesting, diseases weird and fascinating, immunology cool. But cancer, Allison admits, "pissed me off" personally.[36] He was just a kid when he lost his mom to it,[37] had held her hand as she went, not even knowing what the disease was or why she had burns, only knowing she was gone. He'd lose most of his family that way eventually, and though he'd never said it out loud, hadn't even voiced it to himself, in the back of his mind cancer had always been the one potential, practical outcome of his otherwise pure scientific research. And now here he was, with another experiment in mind, and an intellectual path to an emotional destination.

"My lab always has been basic immunology, and half—or less, actually—tumor," Allison said. "But I had a new postdoc [Dana Leach] who had done some tumor stuff. In late summer, I wrote the experiment out. I said, 'I want you to give some mice tumors and then inject them with this [CTLA-4 blocking] antibody. Give others tumors but no anti-CTLA-4, and let's see what happens.'" In November, the postdoc came back with the results: The mice that got anti-CTLA-4 had been cured of cancer. The tumors had disappeared. In the mice that didn't have CTLA-4 blocked, the tumors kept growing.

Allison was stunned—this wasn't what experimental data looked like. "According to the data, it was a 'perfect' experiment, 100 percent alive versus 100 percent dead. Jesus, I mean, I was expecting—something. But this was 100 percent. Either we'd just cured cancer, or we'd really screwed up."

He needed to do it over again. "We had to—it was Thanksgiving, and these experiments take a couple of months." But Allison says his postdoc wasn't going to give up his European trip over Christmas break, not for a bunch of mice.

Allison told him to just set up the experiment again. "Right now, inject all the mice, then go do whatever you're going to do." To ensure that his observations were as unbiased as possible, he told the postdoc to label the cages A, B, C, D. "I'll measure the mice. Don't tell me anything," he said. Allison would do the grunt work and check the results for each, but until it was over, he wouldn't know which group was which.

"It was really harrowing," Allison remembers. He'd come in every day and see that the tumors in cage A seemed to be getting bigger. He'd measure each tumor with calipers and mark the results on his gridded paper, then move to cage B and find the same thing, mice with growing tumors. Same story in cage C and cage D. There were a lot of mice, a lot of numbers, and they were all on the same track. It was 100 percent failure.

Had his break-happy postdoc screwed up this experiment too? Allison felt he was moving backward. Finally, on Christmas Eve he was in the lab, staring at four cages of mice, all with steadily growing tumors. "I said, 'Fuck—I'm not going to measure these anymore. I need to take a break from this.'"

He returned four days later to discover that the situation in the cages had changed dramatically. In two of the cages the mice tumors were now shrinking. In the other two cages the tumors continued to grow. When he unblinded the experiment's cages, he was sure. It had taken time for the immune response to kick in, much like it does with a vaccination, but it had happened. Day by day, and surprisingly quickly, the trend continued; it was just as before—100 percent, a perfect experiment.

He hadn't known where he was going with all of this experimentation, but now, suddenly, they had arrived. They'd figured out a biological mechanism that made sense of decades of confusing data. Tumors learned to express CTLA-4. In mouse models, this was how cancer shut down an immune response. It was evolution, cancer's survival trick, or one of them. If Allison could block it in mice, maybe he could block it in people. The breakthrough wasn't what was in the cages; it was the new view of the world the data revealed. It doesn't usually happen in science like it does in the movies, the eureka moment, a new understanding in an instant. But this was it. *EUREKA!* T cells recognized cancer, cancer shut down T cells, and you could block that.

What else was possible? That question, and the hope it engendered—that was what mattered. And that was the breakthrough.

∼

CTLA-4 had turned out to be what would be called a "checkpoint" on T cell activation, a built-in kill switch poking from the T cell surface, installed by Mother Nature to prevent the body's cell killer from running wild. Allison had discovered it had been hijacked by cancer to shut down (or "down-regulate") an immune response against it.

Allison's lab had made an antibody that found and fit to the CTLA-4 receptor like a key broken off in a lock. It blocked that checkpoint, so cancer couldn't use it. Some biologists compare the action of this checkpoint inhibitor to wedging a brick underneath the brake pedal of a running car.

Checkpoint inhibition differed from previous attempts at a successful cancer immunotherapy that sought to induce, ramp up, or "boost" an immune response to cancer. Instead, blocking the checkpoint prevented cancer from shutting down the natural immune response against it.

For decades, researchers had been looking for something to explain why they couldn't make an immunotherapy that worked reliably against cancer. Many assumed that the problem was that T cells couldn't really recognize tumor antigens—which meant that the problem with cancer immunotherapy was that it was futile to begin with. Work in Allison's lab suggested a different scenario. The T cell could see cancer, but the CTLA-4 molecule acted like a brake, a checkpoint that stopped immune response. Blocking or inhibiting that checkpoint with an antibody might be the missing puzzle piece cancer immunologists had been searching for.[38]

Allison's lab[39] now had antibodies that blocked the CTLA-4 receptor in T cells. They believed they could block cancer cells before cancer got a chance to shut down T cell activation; in theory, this was a potential drug that might help cancer patients. In order to realize that potential, to even know if it worked, it would need to be tested. And in order to test it at scale, it would first need to be manufactured. But Allison couldn't find a pharmaceutical company that was interested.

One problem was that it was 1996 and he wasn't hawking the sort of drug most pharma manufacturers were equipped to make. The easiest, the most common kind, were small molecules. They're relatively simple to assemble in quantity, and the manufacturing process is far more straightforward than that required for the large antibody Jim Allison had for blocking CTLA-4. Most cancer drugs were small-molecule drugs. They didn't cure cancer, but they attacked it, for a while. "That was what was driving pharma then," Krummel says. "And it would be for the next fifteen years."

The other problem was that, while anti-CTLA-4 was a cancer drug, it was one that represented a treatment philosophy that acted not on cancer but on the immune system, unleashing it so it could do its work.

It was, in other words, a cancer immunotherapy. And cancer immunotherapies had heretofore proven to be a risky bet. Manufacturing, testing, marketing, and distributing such a drug (or any drug) would take many millions of dollars and many years. It was a bigger risk than most companies were willing or could afford to take, especially for an approach to cancer most oncologists distrusted.

And as Allison now also discovered, he had a third problem. In the years between when CTLA-4 was first discovered and Allison's and Bluestone's labs figured out how it worked and what it did, a provisional patent had been filed by the pharmaceutical giant Bristol-Myers Squibb. Their patent preceded Allison's as a stake in the ground, but it was based on a misunderstanding of how CTLA-4 worked.

The BMS patent had CTLA-4 as a gas pedal. It claimed their antibody would bind to CTLA-4 as an agonist, revving up the T cell. Allison and Bluestone's breakthrough realization was that CTLA-4 was in fact a brake pedal, down-regulating immune activation. Allison's unique patent was for an antibody that blocked that brake, as a drug for use against cancer. Allison had been right, Bristol-Myers Squibb had it wrong. Allison and his postdocs would eventually prevail. But in the meantime, a conflicting claim against a billion-dollar corporation didn't help their sales pitch any.

"There was all this excitement, and then it was like radio silence," says Krummel. "You could hear the bees in the orchards."

It took two years of travel and talk before they finally found a home, with a small New Jersey–based pharma company created by a team of immunologists from Dartmouth Medical School.[40] Mederex wasn't big, they didn't have the deep pockets of a Bristol-Myers Squibb or a Roche, but they did have a mouse genetically engineered to make human antibodies (rather than mouse antibodies).[41] With Allison's intellectual property, their mice would become living pharmaceutical factories, capable of producing anti-CTLA-4 in quantities sufficient for the first-in-human clinical trials. It might even become a cancer drug, and help people. But that *maybe* was still fifteen years away. Far more likely was that they'd end up curing cancer in mice, one more time.

Chapter Five

The Three *E*'s

If you change the way you look at things, the things you look at change.

—Max Planck

The new discovery was that CTLA-4 was a checkpoint on T cells, a brake that prevented immune activation. Blocking that checkpoint blocked the brake. Allison had found that doing this seemed to change the way the T cell reacted to cancer, at least in mice.

The larger suggestion was that a checkpoint on the T cell might be an important, previously missing piece of a full and successful immune system response to cancer and perhaps other diseases. That response had many critical players, but the T cell was the action star when it came to cancer killing. Most cancer immunotherapy approaches had been trying, and largely failing, to get that star to act. Rosenberg, Greenberg, and others had hoped that enhancing the T cells' energy and raw numbers with the cytokine IL-2 would do the trick. Cancer vaccines attempted to motivate their star into action by introducing the T cells to the distinctive proteins from the cancer cells they were supposed to target and kill. Those approaches had in common a single scientific premise: T cells *could* recognize tumors as non-self and, when they did, fly into action, multiplying and attacking cancer cells. That didn't seem to regularly happen against cancer, and it didn't with IL-2 or the vaccines or other attempts. For years the question had been, why not?

"I found it," Allison would tell me later, referring to both what seemed to be the answer to that question generally, and one of the checkpoints that stood in the way of T cells activating against and attacking cancer specifically.[1] "But I didn't prove it."

That proof was slowly cooking in the labs of other scientists laboring in the trenches at the intersection of the immune system and cancer. Their experiments didn't have anything to do with Allison personally, or anything to do with CTLA-4 at all, but they connected what Allison had just done with that checkpoint molecule at Berkeley into a larger biological story written over hundreds of millions of years. They took a discovery, a *what*, and turned it into a *so what*, and a *why*. Allison had found a distinctive puzzle piece; these other scientists simultaneously uncovered an evolutionary picture missing exactly that shape.

~

Talk to anyone working in the cancer immunotherapy field during the past half century and there are certain names you'll hear again and again, few as reverently as Dr. Lloyd Old.[2] Old was essentially the face and voice of cancer immunology during its darkest hours, a highly trained immunologist and respected academic and researcher based at New York's Memorial Sloan Kettering Cancer Center who straddled the worlds of reputable science and its redheaded stepchild.[3] Old hadn't exactly picked up the mantle of Coley, but with work partly supported by the Cancer Research Institute set up by Coley's daughter, Helen, he honored and improved upon Coley's unarticulated premise, and until his death in 2011, remained a torch-bearing cancer immunologist of the first order.[4] For fifty years he gently brought new talent into the field, helping standardize and improve rational scientific approaches to testing various strategies weaponizing the immune system against cancer.[5]

Many of those approaches involved a tumor strain he had created, called "Meth A." This was a model tumor he could experiment with, and pit against various proteins possibly related to immune response.

One of those proteins was a chemical messenger he'd identified as a potentially important factor for tumor killing, which earned it the name tumor necrosis factor, or TNF. We now know that TNF is a cytokine, one of the dozens of powerful chemical alarms that initiate specific steps in immune response to disease. TNF is part of an instruction that a T cell gives a cell it has targeted for death. The instruction is for that cell to please neatly kill itself.

Our bodies are constantly shedding old or damaged cells, allowing new cells to take their place. This natural process of cellular self-destruction (called *apoptosis*, from an ancient Greek word meaning "falling off") is hardwired into cells. The process is a cellular spring cleaning. Over the course of a year, each of us will shed a mass of self-destructed cells approximately equal to our full body weight. The body uses this natural process to rid itself of cells that are damaged, infected, or mutated. Even before we're born, apoptosis plays a vital role in the early stages of infant development in the womb. Some mutations that result in cancer disable the self-destruction capability of apoptosis so that instead of self-destructing and being replaced by a healthy cell, the mutants keep dividing and multiplying out of control. Resistance to apoptosis is one of the critical so-called hallmarks of cancer. Old's experiments were aimed at better understanding it, and trying to short-circuit that short circuit.

TNF seemed to be involved in that apoptosis process. Old had found that in his mouse models, adding additional TNF would induce the immune system to destroy the cells of the Meth A tumor. It was something like a cancer research version of destroying a ship in a bottle.

But to really understand the role of TNF in immune response, he also needed to remove the cytokine from the system—to remove, or block, that link in the chain, and see what happened, or didn't happen.

Half a country away, Dr. Robert Schreiber[6] was running just such a test. A Rochester, New York, native and immunology research head at St. Louis's Washington University School of Medicine, Schreiber wasn't trying to make the immune system do anything specifically. He certainly wasn't trying to make it do anything with cancer. Like his former colleague Jim Allison (the two men had crossed paths as Scripps postdocs), he just wanted to understand as much about the immune system as possible, pushing the science forward question by question, inch by inch. This was complex and relatively unexplored territory. Such questions were more than sufficient to last a lifetime.

At the time, Schreiber's lab had fifteen staffers looking at the immune system's messenger chemicals, the cytokines. Several had been identified, and each seemed to be involved in the complex dance of signals, actions, and reactions. Schreiber's lab had developed a number of proprietary antibodies, ingeniously grown in Armenian hamsters.[7] Each antibody matched up to a specific cytokine. Each had proved to be especially good at blocking that one signal without affecting the others. Each removed one of the links in the larger chain reaction of immune response.[8] They were doing good work, leading the field in this particular aspect, and Schreiber was perfectly happy to keep doing it. It was, he says, "a lot of fun."

Then on an unassuming spring Tuesday in 1988,[9] Schreiber

got a call from Lloyd Old that changed the direction of his work and life.

It started off innocuously enough. Old had need of good antibodies that blocked TNF. He wondered if Bob Schreiber wouldn't mind lending him some. "So I said sure," Schreiber recalled. In fact, if Old was interested, Bob would send over a whole menu of antibodies; they had plenty. Yes, they were proprietary to his laboratory, but this was science. It was like lending a neighbor a cup of sugar.

Schreiber's lab pipetted the antibodies into test tubes, packed the racks in liquid nitrogen, and overnighted them to Old's lab. Not long after, Old was on the phone again. He sounded excited. "Lloyd said the anti-TNF antibodies we'd sent had worked very well," Schreiber remembers. They'd targeted the TNF and muted the cytokine's battle cry to the immune system. It didn't stop all immune response in mice, but it quieted it significantly.

Old had also tested out some of the other antibodies Schreiber sent over, molecules that blocked other immune cytokine signals. The one that blocked the cytokine interferon gamma (IFNγ) worked the best, shutting down TNF immune response even better than the one directly blocking TNF, which was surprising. Shutting down IFNγ pretty much completely shut down the immune response against Old's tumor cells.

"So Lloyd said, 'How do you think that works?'" Schreiber remembers. That turned out to be a gateway question.

The answer would require more experiments, moving from tumor cells in a test tube to tumor cells transplanted into mice at Bob Schreiber's lab. Next thing he knew, Schreiber was neck deep in the project.[10] Lloyd Old was an old hand at drawing smart young biologists into his cancer immunotherapy quest.

Schreiber's lab grew Old's Meth A tumor cells and transplanted them into two different groups of mice. This was a variation of the experiment Old had done, asking the same question about interferon gamma, but in a different manner. This time, instead of using antibodies to block the interferon gamma, they used mice with a mutation that rendered their interferon gamma receptors defective.[11] Whatever it was interferon gamma did to the immune system of the other, normal (or "wild-type") mice, it couldn't do it in these mutant mice. And in those mutant mice, Old's transplanted tumors prospered; those mutant mice got cancer. The normal mice, with normal immune systems, didn't.[12]

Schreiber was a pure research guy, and his lab really wasn't interested in cancer immunotherapy,[13] or any therapy at all. "I told Lloyd, I wasn't really much aware of what was happening in the world of tumor immunology," Schreiber says. Old told him that wouldn't be a problem. "So he pretty much taught it to me, over the phone."

Using Lloyd Old's special line of tumor cells, Schreiber had found a way to shut down an immune response against it, by blocking a cytokine fire alarm in specially bred mice. Now then, Old wondered over the lab phone, what did Bob think—would those mutant mice be more or less likely to develop *real* cancer? Not transplanted cancer from transplanted model tumors, but real homegrown cancer that arose from mutations of their own body cells?

By accident or design, the questions Old was asking were steadily marching them toward the edges of the intellectual battleground where Old had spent his career. Schreiber didn't know it, was totally oblivious to the war being waged around cancer immunotherapy, and had no clue that he was about to steer his prestigious lab into an academic minefield.

In fact, he already had. Though it wasn't his intention, Dr. Bob

Schreiber's work had just weighed in on the foundational question in cancer immunology: whether cancer and the immune system had some interrelation.[14]

The notion that they did had been posited in 1909 by the German physician and scientist Paul Ehrlich.[15] Ehrlich predicted that the immune system looked over and protected us from most of our mutated body cells much as it did everything else foreign to the body, and that without such "immune surveillance," cancers would be far more frequent. Fifty years later a more detailed understanding of tumor biology and the rejection of organ transplants gave Ehrlich's concept new life and new champions in the Nobel Prize–winning Australian virologist F. Macfarlane Burnet, the American writer and physician Lewis Thomas, and Lloyd Old, among others. But the theory was severely undermined by the stark fact that cancer patients had nothing to show for it.

And so while cancer immunologists like Old continued to insist that cancer and the immune system had something to say to each other, most people firmly believed that they did not. For proof they needed only to point at the clinical experience of nearly every cancer specialist and immunologist in the world, and an experiment at Memorial Sloan Kettering on nude mice.[16]

∾

Nude mice are lab animals bred with a genetic mutation that renders them completely hairless. (Mice are born that way, pink and smooth as pinkies; nude mice remain this way throughout their life spans.) That was handy for scientists, because it made them easy to distinguish from normal mice; in addition to lacking fur, these mutants were also born without a thymus,[17] the tiny butterfly-shaped organ where T cells mature.[18] No thymus, no T cells—and, it was assumed, no adaptive immune response.

In 1974 a Memorial Sloan Kettering Cancer Center researcher named Dr. Osias Stutman injected two colonies of mice—one nude, one wild-type (normal, with normal immune systems)—from syringes loaded with a whopping dose of a highly carcinogenic stuff called 3-methylcholanthrene. If T cells conducted helpful "surveillance" and policed cells for mutations (as recognized by their unique antigens), mice without immune systems should get more cancer, faster and worse than their wild-type cousins. Instead, Stutman found that both mouse populations quickly developed tumors,[19] which grew at the same rate and quantity. There was no difference between the experimental cages, and thus, the experiment suggested, no such thing as immune surveillance of cancer, and, by extension, no point in trying to unleash the immune system as a defense against it.

Stutman's paper had landed like a bomb in the definitive peer-reviewed scientific journal, *Nature*.[20] The implication was that the small but persistent cancer immunotherapy field was a dead end. For true believers like Old and others, the findings had to be in error, experimental results based on a mistaken experimental premise—a phenomenon researchers refer to as "garbage in, garbage out." But at the time nobody could identify that mistaken premise. In terms of the opinion—and research funding—for cancer immunotherapy by the scientific community at large, the Stutman paper had been, as the *British Journal of Cancer* put it, "devastating."

Schreiber wasn't taking aim at the experiment's misunderstanding of nude mice, or the other hidden flaws within Stutman's experiment.[21] He wasn't even aware of them. He was simply testing the rather interesting question Old had so charmingly asked.[22]

That test started with two populations of normal wild-type mice. In half, they injected Schreiber's antibodies, which blocked off the "seek and destroy" cytokine signaling to their adaptive immune response. The other mice were left alone, with their immune systems intact. Then both populations of mice were injected with a drug that encouraged their cells to mutate and grow tumors.[23]

"The immune-suppressed animals developed more tumors more rapidly than the normal animals," Schreiber says. "That was really interesting." It had been a good experiment. Now it was time to share the findings. At the time, he truly believed it would be that simple.

∼

Each week the university's laboratory chiefs gathered in a conference room to share updates on their experiments and investigations. Schreiber was excited. He had something new and interesting to share, and he expected there would be a lot of questions. What he didn't expect were rebuttals. "There are no danger signals in tumors," one chief said. "Cancer cells are too close to normal cells to be recognized as non-self," said another, "so they are not subject to immune notice."

Schreiber couldn't believe it—he had the data. Data was data, the foundation of science. But his colleagues were ignoring the data, even as they became surprisingly, personally indignant at its implications. In short, Schreiber felt they were reacting as if he were talking religion, rather than empirical science. And the most surprising thing was that he knew these people. These were excellent scientists, his peers and colleagues. Many were his friends. If his close academic circle reacted this way, Schreiber wondered, how would the outside world react? "That was my first inkling," Schreiber says. "I saw now that we were getting into, uh—something differ-ent, when it came to cancer."

Scientists are people; they have beliefs, and are personally invested in them. And that can sometimes lead to unintentional and often unconscious bias, and a sort of intellectual blindness. It can, in other words, make even scientists unscientific. Schreiber sent a paper to the leading scholarly research journals detailing his findings. He had good, hard data suggesting that blocking this cytokine in mice had made them more susceptible to cancer. "And I was really surprised by the responses," Schreiber says. "They'd say things like, 'Oh, you're trying to say that there's cancer immune surveillance. Don't you know that cancer immune surveillance doesn't exist?'"

In fact, Bob wasn't trying to "say" anything—he was present-ing data, doing science. "It happened again and again," Bob said. "I kept saying, 'But look at the data. The data was so clean!'"[24]

Bob Schreiber is an exceptionally mild-mannered guy, gentle even. But this, he admits softly, "became very frustrating."

"Here we had developed beautiful data, and people negated the science by saying, 'I don't *believe* that the immune system can see the tumor.'" Schreiber's take was that the cancer biology commu-nity was so dead certain that immunology was a waste of time that none of them had looked at the biology of the tumors that closely. Schreiber and Old had looked and seen something new, but now nobody was willing to accept their findings, becase they conflicted with their prejudices.

Finally they realized the only way forward was to overwhelm the ignorance with a tsunami of more: more experiments, more mice, and much, much more data, data so big and beautiful and clean that "even the most critical reviewer would have to accept the papers."

It took three additional years. The data was big and beautiful. And even then it didn't work. The big scientific journals (Schreiber is too polite to name names) still wouldn't touch their findings.

This time he wasn't surprised at the intellectual fireworks, though he still cringes a bit remembering some of the heated exchanges at scientific conferences. "It got to the point that people actually started looking forward to the conflict," he laughs. His good work had become a sideshow, but at least it wasn't being ignored. "And actually, I think that brought more attention to the findings."

But rather than continue to repeat the experiments that might hammer and "win" the current debate, Schreiber and Old moved on. In the course of their three-year tsunami of experiments, other interesting observations about cancer and the immune system had emerged.

Immune-blocked mice made a lot of tumors quickly, but the tumor cells were weak and simple. When those tumors were transplanted into immune-normal mice, their immune systems quickly recognized and killed those tumors.

The opposite happened when "normal" tumors that grew in normal mice were transplanted into mice without immune systems. This time, the tumors grew like weeds and quickly killed the mouse.

"And that," Schreiber says, "was when the lightbulb went on."

Immune surveillance by T cells spotted and killed most mutated cells long before they posed any risk of becoming what we'd recognize as cancer. As a result of this rigorous surveillance, all that were left were cancer cells with advantageous mutations that helped them to survive and grow. Comparing the tumors from his two mouse populations, Schreiber discovered that the tumors on mice without an immune system were simple, obvious, and defenseless—the weak wildebeests that, in a world without predators, had been free to reproduce and swell into a herd. But in mice with intact immune systems, what Schreiber and Old saw was essentially a stronger herd: stronger, fitter cancer. It seemed the evolutionary pressure acted like a sort of editor, drawing a red line through only the most obvious or defenseless cancer cells, ensuring that only elegant, powerful, and tricky killer sequences survived.

Previously, the debate about immune surveillance was a yes/no question—did it happen, or didn't it? But these experiments suggested a more complicated relationship between the immune system and this disease—a dance in which each partner reacted to the other and was changed. The scientists wrote up the data and named the phenomenon "immunoediting,"[25] and sent it in to the journals again. Finally, the data tsunami could not be ignored. This paper found a more accepting audience, including the editors of the prestigious scientific journal *Nature*.

They now had a potential new model of how cancer worked. "We had defined the first and the last part of the process," Schreiber says. The first part was the immune system killing cancer as it developed—what they called "elimination."

The second part involved the mutated cells that the immune system didn't kill off—they were still there, and still mutating. These cells that got away from the immune system might become cancer as we know it, the difficult, tricky, and deadly product of those cells fit enough to "escape."[26]

Meanwhile, years of gathering his data tsunami had left Schreiber with a lot of mice. "They were just there, eating up my grant funds," he says. He figured he should do something with them.

These were the normal immune mice; none of these mice had tumors, even though they had been subjected to serious carcinogens. Bob wondered if maybe the tumors were being held in some sort of check or a dormant state by their immune systems. All that was required to find the answer was to block their immune systems, as he had the other mice, and see what happened.

"And sure enough, many of the mice that we thought were cancer-free developed tumors extremely rapidly."[27] And when they tested those tumors they discovered that they were "obvious" tumors—they had no tricks to hide from the immune system. Most had strong mutations that expressed a protein (antigen) that a normal immune system would easily recognize as foreign and kill.[28] These tumors had been there all along, dormant and invisible, held in check by the immune system but not totally killed by it either. And yet somehow they had survived in these mice by holding the immune system "in check" as well. How had they done it? How did the tumor cells survive?

Schreiber was aware that Jim Allison's lab had already found one answer—a checkpoint. While the immune system was activated against the tumor, it eliminated the weakest of the herd—the cells with the most obviously foreign antigens, the ones the T cells recognized. But while the immune system was busy killing the weaker mutations, some surviving mutant cells were able to change their genes, and express something on their cell surfaces that called off the T cell attack. Those "stop" signals the tumors now expressed activated checkpoints like CTLA-4 on the T cells; the cancer had evolved and learned to pull the hand brake on the killing machines and survive. Old's and Schreiber's experiments helped explain the story behind the checkpoint Allison's antibodies blocked, and how the checkpoint made sense of their observations of a disease that is at once elusive and prevalent, and confoundingly given to remission.

The implications were that some cancers might remain held in a sort of "equilibrium" with the immune system for years, perhaps even a lifetime. Some cancerous mutations would be recognized and attacked by the immune system, but that just thinned the herd of its weakest members and provided other cancer cells a chance to call off the attack, mutate further, and regroup. Eventually, given time or opportunity, those surviving cells might find a way to escape from the equilibrium. Cancer and T cells responded to each other, reacting and changing in an elaborate immune dance.

This mirrored some of what oncologists saw in their clinics, such as cancer patients who exhibited sudden remissions after having had no evidence of the disease for sixteen years, or the cancers that emerged in patients with weakened or compromised immune systems from age or chronic inflammation or immune-compromising disease. It also potentially helped explain why a cancer that returned often no longer responded to the previously successful therapy.

Seven years later, a more nuanced version of the three E's— elimination, equilibrium, and escape—helps redefine our understanding of the relationship between some cancers and the immune system.[29] Immunoediting describes how the immune system protects and defends the host from cancer, even as it "sculpts" the genetics of some tumors. Those tumors escaped because they had successfully evolved tricks to evade or shut down the immune system—and some of those tricks included taking advantage of the safety checkpoints built into the T cell. Allison had found one;

soon other critical checkpoints would be discovered. Blocking those checkpoints, the work suggested, might thwart an important survival mechanism cancer had developed, and let the immune system do its job.[30] This was the basis of what Allison was testing in clinical trials. When Schreiber and Old published their three *E*'s paper in 2004,[31] it was clear the theory dovetailed with what Allison had found in 1996. The men had essentially come to the same place from two different directions; in 2017 Allison and Schreiber would share the prestigious Balzan Prize for their complimentary work.

Jim Allison puts it all much plainer and in a south Texas accent. "Yeah, I found it," he says, "but it was Bob proved it."

Schreiber and Old published their paper just as researchers were making even more breakthrough discoveries about the tricks cancer uses to shut down the immune system. Immune editing lives on as an important concept, but its interpretation has changed to emphasize the revelation of the role played by immune checkpoints. The current understanding, perhaps best illustrated by the 2014 "cancer immunity cycle,"[32] is that while most cancer cells express antigens that T cells can recognize and attack, cancer has evolved to manipulate T cell checkpoints, as well as other tricks, in order to shut down the immune system and survive; if it didn't, cancer wouldn't exist. Those tricks can be thought of as a sort of fourth E: Emergency kill switches on T cells (including checkpoints, such as CTLA-4) that cancer cells have Evolved to Exploit. This model helps explain why the immune system—and cancer immunotherapies—had previously failed to kill cancer. As long as cancer could exploit the checkpoints and shut down T cells, immune defense was futile. But now some of those exploits were exposed. And they could be blocked. Which opened the possibility for a hopeful fifth and final E: the End of cancer.

~

By 2001 Allison's drug, an anti-CTLA-4 antibody he'd patented as 10D1, was in production by Mederex[33] as experimental drug MDX-010.[34] Finally it could begin the long climb toward possible FDA approval. The first step was to ensure that the drug was even safe enough to test. It was tolerated in small doses by macaque monkeys without significant toxic effect,[35] then given to human patients at the same dosage, a one-time injection of 3mg of MDX-010 anti-CTLA-4 antibody, administered at a private cancer clinic near UCLA. Only nine patients participated, a brave and desperate lot of stage 4 metastatic melanoma volunteers with no other treatment option, each willing to contribute to the science and all hopeful to survive. Even at these low doses, seven of the patients developed rashes and other adverse effects consistent with unleashed immune response; none were bad enough to stop MDX-010 from moving to the NIH for its first true clinical trial.[36]

In early 2003, Jim Allison packed a winter jacket and traveled to Bethesda, Maryland, to join Dr. Steven Rosenberg's team at the National Institutes of Health laboratories for a phase 1 clinical trial to test his experimental drug for safety in humans. Once again, a small but desperate group of melanoma patients competed to fill the twenty-one slots;[37] almost all of them had already tried an earlier form of immunotherapy, including IL-2, or interferon, and almost half had already tried chemo. The drug was injected over ninety minutes every three weeks. And again, MDX-010 passed the safety hurdle; three patients responded to the drug, two of them completely, and one patient even saw her tumors dramatically disappear during the study.[38]

Results from the others had showed some tumors shrinking, some growing, and some not responding at all during the twelve-week course.[39] Several suffered toxic reactions, essentially autoimmune events serious enough to require ICU visits. Clinicians managed to wrestle those cytotoxic events under control,[40] and the results were sufficient to proceed to a phase 2 trial. The hope was that more patients, bigger doses, and additional time would demonstrate statistically meaningful positive results, with sufficiently low toxic events. With good enough phase 2 results, they might even be able to skip the final stage 3 trials and go straight to approval as a cancer medicine.

The goalposts of success for these next trials would be relatively modest, as new drugs go. The anti-CLTA-4 antibody was being tried out as a candidate for a "last line" therapy for specific melanomas. *Last line* refers to the drug you turn to when everything else has failed. Allison had his suspicions about highly mutated, and thus highly antigenic, tumors being the ones most likely to rely on something like CTLA-4 to be able to survive the immune system.[41] Melanoma, with its high mutational profile and low late-stage cure rate, seemed a good potential target for this hopeful new drug.

Melanoma isn't the most common form of skin cancer, but it was the most deadly, responsible for three-fourths of all skin cancer deaths. In 2000, 75 percent of the patients who received a stage 4 "skin" diagnosis would not survive the year. Ninety percent wouldn't survive five. One of the reasons this disease was the first target for testing this new immunotherapy was simply a lack of better options.

~

Melanoma is a particularly tricky and aggressive cancer resulting from highly mutated skin cells. It moves fast[42] and keeps mutating.[43] As a result, metastatic melanoma patients often moved from drug to drug, trying to keep up with the rapidly mutating cancer cells, a frustrating and usually futile exercise that melanoma specialists like Dr. Jedd Wolchok had seen over and over again in his cancer clinic at Memorial

Sloan Kettering Cancer Center.[44] "Nothing much worked against metastatic melanoma," Wolchok says. Cancer is rarely a good-news business, but melanoma was especially grim. Now, as a primary investigator hosting a cohort of patients to test Allison's drug in phase 2 trials, Wolchok hoped to see something better.

Wolchok has a boyish face and a gentle demeanor that makes it hard to imagine he's been working in immune oncology longer than almost anyone of his generation.[45] Like Dan Chen and other MD-PhD immunotherapists with both a lab and a clinical cancer practice, Wolchok had gotten plenty of resistance about his beliefs from his chemotherapist peers over the years. But he nevertheless continued to have faith that an immune approach to beating cancer could work. Thirty years old at the time, Jedd Wolchok was too young to be considered an immunology true believer, but he had an equally deep conviction born from his experiences as a high school summer intern at Memorial. He had witnessed a patient's full remission after receiving an experimental cancer immunotherapy vaccine. That experience, quickly followed by a personal introduction to the legendary Lloyd Old, had pretty much sealed Wolchok's career path. The course of his career[46] and the steady stream of papers coming out of labs at the NIH[47] had only deepened his conviction that, under certain conditions, T cells could be activated against human tumor antigens, and attack and kill cancer.[48]

Wolchok had been following the papers coming out of Allison's lab too, and he wondered if the Texan hadn't just redefined those "certain conditions." CTLA-4 seemed to be a missing link—the brake they'd never known about, but that had been on the whole time. That would explain a lot about what Wolchok was seeing in the lab and clinic, where he was working on vaccines.[49]

Wolchok had put his hand up early to get involved in the anti-CTLA-4 drug trial. Now he was both excited and nervous. He'd heard about some promising responses and also heard the horror stories coming back from the phase 1 investigators. "They were really emphatic about the side effects," Wolchok remembers. They kept repeating their warning like shell-shocked survivors: *Really—the side effects need to be taken very seriously*. This was new territory for everyone, but already the message was that the immune system without brakes could be harrowing, and potentially deadly.

~

Before these trials kicked off, Mederex was bought by the pharmaceutical behemoth Bristol-Myers Squibb. BMS had the deep pockets required to fund what was essentially a risky bet, and just as critically, they had the resources to survive a failure, should it come to that.

Time is money in clinical trials, and the goal is either to fail or get to market as quickly as possible. Usually drugs go through three phases of clinical trial. Each phase takes years and costs many millions of dollars. Not surprisingly, the new owners at BMS were hopeful to fast-track their new drug through the costly FDA approval process if possible.

Traditionally, new cancer drugs are approved on the basis of having demonstrated better results than the previous standard drug for patients, as defined by a set of rules agreed to beforehand. These rules are the endpoints of a study—essentially, the goalposts. Bristol-Myers Squibb proposed a deal that changed the endpoints in order to shorten the game, offering an unambiguous standard of success that would make a phase 3 trial unnecessary. After all, they weren't trying to supplant any of the traditional chemotherapies; they only wanted to be the last line for a type of cancer in which most patients failed.

So BMS offered the FDA a deal. The goalposts for a "good response rate" would be 30 to 50 percent of their patients having their tumors shrink during the three-month trial. If they succeeded the drug would be approved, skipping phase 3 and going straight to market. If they failed, they failed faster. At the time, BMS was so certain that they would meet those endpoints that they failed to recognize the corner they were boxing themselves into.[50] They didn't understand how an immune drug worked—basically nobody did, because an effective one didn't exist. Instead of allowing for that uncertainty, they'd arranged for their breakthrough drug to be judged like a traditional chemotherapy. The FDA agreed. BMS celebrated a coup. And the study failed. Axel Hoos knew it would.

~

Axel Hoos is an MD-PhD who had trained in his native Germany. Hoos is precise in manner and unmistakably Teutonic in accent, a lean, bespectacled man with a close-cropped blond crew cut and a crisp wardrobe of suits. As an MD surgeon who became a PhD immunologist after he finished his specialty training in surgery, molecular pathology, and tumor immunology at Memorial Sloan Kettering in New York, followed by business school at Harvard, Hoos is another of the strangely prescient characters in immunotherapy who positioned themselves perfectly to catch a falling star.[51] As a group, cancer immunologists trained hard to get lucky.

Hoos was scanning the horizon for a drug candidate capable of changing everything when Jim Allison and his anti-CTLA-4 antibody showed up on his radar. Hoos had watched Allison license the antibody to Mederex, which handed the drug to Bristol-Myers Squibb for phase 2 trials. This was exactly what he was looking for: the (maybe) right drug, in the hands of a company with the right resources to see it to success. Hoos applied and became global medical lead for the BMS immuno-oncology program—the pinchpoint between the suits and the white coats, the guy in charge of all the studies of a potential breakthrough drug.[52] He'd timed it perfectly, only to discover there was a catch.

∾

For as long as the FDA has been in the approval and consumer-safety business, cancer therapy always had the tumor as the target.[53] When those drugs work, the tumors shrink. When they don't work, the cancer continues "progressing."[54]

The endpoints of medical studies—and the language describing those endpoints—are both important. New cancer drugs were evaluated by whether they delayed progression longer than the old drug. That gift of time is called progression-free survival, or PFS, and was the standard measure for evaluating any new cancer therapy.

Survival is important. So is feeling better. Cancer patients—all of us—want both. But "feeling" is subjective, and harder to standardize and quantify than progression, which is a physical measure of the shadows on a CAT scan.[55]

Cancer immunotherapy was new, unproven, and largely untested. It was also different than all previous cancer therapies. It worked on the immune system, not the tumor directly, and nobody knew what that would look like. Describing it in terms of PFS contained an assumption that a new, unfamiliar mechanism of action would somehow look like the old. In the end, this habit proved to be an invisible bias. It was a way of thinking that gave this first generation of cancer immunotherapy drugs no chance of succeeding.[56]

∾

Most researchers had never seen a working immune cancer therapy before. But the women and men who ran the clinical studies were now discovering that immune therapy doesn't look like targeted therapy, or surgery or radiation or chemo. And its numbers look different, too.[57]

Over the short course of the study, patients' PFS graphs were wavy lines, or zigzags. The tumors would swell, shrink, then swell again. Either the drug wasn't working, or it was working differently than chemo did. Either way, it looked like a failure on paper.

But in the clinic Jedd Wolchok saw something different. While several of his patients looked worse on the scans, they reported feeling better. Some had tumors getting bigger. Some had tumors melt away in one place even as new tumors appeared in another.

Cancer immunotherapies live at the intersection of two living systems: that of the tumor, and the immune system. Scientists had long known that autoimmune diseases such as Crohn's disease don't progress in a linear fashion. Scans and tests look like reports from a battle ebbing and flowing between two living forces, self versus self. The anti-CTLA-4 results had that ebb and flow. Perhaps, the researchers reasoned, what they were witnessing was an immune response, rather than a tumor response. CTLA-4 caused the immune system to react to cancer much as it would any other infection, in waves of fever and swelling and inflammation. It was an attack, punctuated by safety checks of *Are you sure?*

The results of the trial showed that only 5.8 percent of the patients were progression-free. Results like that indicated a cancer therapy hardly worth considering even if it was approved. "And so the FDA said, 'Go away, we don't want this,'" Hoos says. From a data perspective, Jim Allison's new miracle drug really hadn't improved much on Coley's Toxins. By many measures it was far worse.

Most pharmaceutical manufacturers spread their bets across a large number of potential drugs, hoping at least one will work and pay off the debt incurred by the others. They didn't need to bat a thousand to succeed, only to be right more often than they were wrong. In this business model, scrapping anti-CTLA-4 would be the price of doing business.

At the same time, the Swiss pharmaceutical giant Pfizer was also testing its own anti-CLTA-4 antibody. Hoos felt that the two versions were, in biological mechanism, the same; the only differences were the dosages they'd be tested at and who owned the data. The two trials began at almost the same time, and the first readouts were essentially the same for both.

Pfizer saw the numbers and scrapped their study. Hoos was determined to continue. BMS had invested a lot. And the science and logic were there. There was impeccable data that provided evidence that immunotherapy can work against cancer.

"Whenever something new happens, you have to break old habits and old thinking," Hoos says. "And for many people, that was hard."

Axel Hoos picked up the phone and dialed Wolchok. He was one of the bright young immunotherapy oncologists out there, and he had his heart and soul in these studies. They all did. Hoos wanted them to know he was talking to the committee. "We're not going to abandon the program because of these phase 2 trials," he promised. "We're not going to let this go."

Hoos planned to fight, and he wanted to be sure Wolchok and the other investigators were ready to join him.

"That call was very important to us," Wolchok says. "Because in the clinic, we were seeing very unusual things.[58] There were some miraculous stories buried in the data." In Wolchok's lab, one of those miraculous stories belonged to Sharon Belvin, another to a Mr. Homer.[59]

Sharon Belvin's visit to Wolchok's MDX-010 trial wasn't her first stop at Memorial Sloan Kettering. She'd already been around the block with several cancer therapies, including chemotherapy and IL-2. Some had succeeded for a while, all had ultimately failed, and at twenty-four years old, she was, by Wolchok's estimate, only weeks away from dying. But by her third treatment she was out of her wheelchair, well enough to walk her dog between appointments. She looked noticeably better—the pretty young blond athlete had begun to emerge from the wracked gray shell. The drug seemed to be working for her,[60] and she wasn't the only one.

Mr. Homer was fifty years old and had stage 4 melanoma that had progressed to his kidney, liver, and lymph nodes. It looked bad on the scans, and after twelve weeks of the anti-CTLA-4 antibody, those scans looked worse. Blood tests showed that his lymphocyte count had skyrocketed, suggesting his T cells had been unleashed and were picking up on the cancer, and at the end of the twelve weeks he reported feeling better. But what mattered were the increased number of metastases on his liver and the visibly increased tumor burden in his kidney.

The numbers were the yardstick. But it was the stories that made the case for the breakthrough.

"This is a field that built its foundations on anecdotes," Wolchok says. Traditionally, anecdotes are discounted by researchers. "The plural of *anecdotes* is not *data*," they say. "But sometimes, anecdotes are really important," Wolchok explains. And they're especially important when the biological mechanism being observed is not fully understood.

Something important was happening with this new cancer immunotherapy drug, one that would come to be called ipilimumab, or "Ipi" for short. It wasn't happening in every patient the same way, and it wasn't even happening in every patient, but it was real. A few did see their tumors shrink, some saw them grow, many saw both. "We had patients who had enrolled in the study that were so close to death, they had already been sent to hospice," Wolchok explains. They'd gone through the study, they felt better, but their scans were worse. At the end they were sent home rather than to hospice, negative data on the study's statistical chart.

"The same patients would call back, six months later—'Oh, hey, I'm alive!'" Wolchok says. "And their cancer was gone. The scans proved it."[61] And it wasn't just his team that were making such observations.

Hoos was in the study's data hub. He had the numbers coming in from trials around the world, and he had clinicians reporting back with the same sorts of mixed observations Wolchok had seen.

"We saw some clinical observations that seemed, on the surface, counterintuitive," Hoos recalls, "patients who say they're doing much better, when on the CAT scan it looks much worse. That's because the immune system is sending its army of cells to the tumor, which makes it look bigger on a CAT scan."

In the short run, the patient's feelings proved to be more sensitive than the imaging technology. "The patient had symptomatically improved but the doctor couldn't see it on a CAT scan," Hoos says. "That gave the feeling, there's a lot more here than meets the eye." It also made clear that to accurately assess immunotherapy, they'd need to rely more on the physician's art of observation and less on tumor scans.

In time, they could redesign what a successful cancer immunological test looked like, and rewrite criteria for the FDA. But that future might not come to be if they gave up on this drug.

"You couldn't force new criteria down the FDA's throat," Hoos says. Instead of changing the game, they just needed to move the goalposts.

At the end of the day, the point of PFS, the part that really matters, isn't the P, but the S—not the tumor's progression, but the patient's survival. "It's the best endpoint possible," Hoos says. "If the drug has real benefit, it should make that person live longer."

That might seem obvious, but it wasn't part of the clinical protocol. By the chemo-based criteria, patients who felt better even if their tumors didn't look smaller on the CAT scan were "unconventional responders." They were survivors, but they didn't count.

Hoos believed they had the data to succeed. If you ignored progression of the tumor and focused only on survival of the patient, the statistical curves revealed an important new medicine—and a possible breakthrough in our war against cancer.

Wolchock's patient Mr. Homer was continued on the anti-CTLA-4 drug after his initial twelve-week study period. He'd come in as a candidate for hospice, but at week sixteen he no longer had crushing abdominal pain, and he even felt well enough to go on a short vacation with friends. A year later scans showed that his lesions and tumor had melted away almost entirely. But back in 2006, Mr. Homer didn't count. During the twelve-week window of the anti-CTLA-4 drug study, his tumors did not look smaller on his CT scans. He did not constitute progression-free survival, and so his data argued against approving the drug that ultimately saved his life.

But could they prove it? "We didn't have to convince the FDA of this," Hoos says. "If you can show survival benefits, the FDA doesn't care why it's a survival benefit! As long as it's real. And this was real!"

The more difficult task was to convince Hoos's bosses at BMS to continue the study, extend it, and use new endpoints that measured overall survival.

"When you change your endpoint from progression-free survival to overall survival, your time line gets longer," Hoos said. "Longer by three years!" The cost of a five-hundred-patient trial for three more years might run into many, many millions of dollars—for an experimental drug that had already "failed" by standard measures.

Nevertheless, the company agreed to it.

"We had enough conviction to stay with it. At several points, it could have failed. You do the wrong trial design. You don't get the backing internally. You partner with the wrong group, you measure the wrong endpoint. Or you draw the wrong conclusion, even if you do the right thing. There are a million ways to fail. This is true for the clinical research world everywhere. It's only those who are tenacious enough to go through with the experiment and not get deterred early," Hoos says. "That's where real breakthroughs come from."

This was an exhaustive, careful plan that took millions of dollars and half a dozen years to execute and changed the face of medicine. But Hoos can summarize how to make a breakthrough therapy with Germanic certitude and a single sentence. "So once you have a mechanism that works [CTLA-4], and you have some persistence and conviction, you follow that in the clinic, you create a method that allows you to detect these things properly and make it visible to the FDA and others that need to buy it, then you arrive somewhere," he says with a smile. "And that's in very *very* simple terms the Ipi story."

"The word *cure* can now be used in oncology," Hoos says. "It's no longer fantasy or a cruel promise that you can't fulfill. We don't know yet who will be the lucky patients that will be cured, but we have seen cures already. When we started, in 2011, we started seeing cures for individual patients."

When they first unblinded their longer study, the survival rate of metastatic melanoma had already been improved. "With ipilimumab we were at 20 percent overall survival," he says. "That's quite a step in the right direction, and these numbers are continuing to increase." Those increases are now compounded by using combination therapies; the data continues to emerge, and the numbers change, on a nearly monthly basis. "So for some it's a functional cure, for others it's a true disappearance of all disease and it never comes back, so it could be a true cure," Hoos says.

~

Ipi is not a cure for cancer, but the success of Ipi was the breakthrough of cancer immunotherapy. It lit a fire under the cancer research community and changed the direction of work for the decades to come. Suddenly, years of failed experiments in cancer immunotherapy needed to be reexamined in the light of the news that they'd been trying to drive the immune system with the hand brake on. And for the first time, they might know how to take it off.

This new understanding of the tricks cancer had evolved to escape immune surveillance inspired researchers from fields across the scientific spectrum to enter the immunology field. For those already in it, the race was on to look harder for other checkpoints, and maybe other brakes. Most critically, the breakthrough made it unambiguously clear that the human immune system could be helped to recognize and kill cancer, opening a hopeful new front in our age-old war against the disease.

This was cancer's penicillin moment. We're still in it, and that's exciting. But if the history of cancer immunotherapy teaches us nothing else, it should be that hope needs to be heavily tempered with caution.

Chapter Six

Tempting Fate

We search where there is light.

—Goethe

Every cancer patient's story is a journey, some longer and harder than others. Brad MacMillin's lasted twelve years. It began in 2001, and started with a spot on his heel, a dark circle under the callus, like a bubble in ice. Brad was a jogger and weekend basketball warrior; he'd had blood blisters before, but this one seemed to be getting larger. After his annual checkup he was referred to a dermatologist. The dermatologist wanted to remove it immediately.

The urgency surprised Brad; so did the size of the divot she carved from his foot. She would get it to a lab, he should wait.

Brad found his wife, Emily, in the waiting room, alone in a sea of chairs. It was past five on Friday, Memorial Day weekend. The lab people seemed to staying just for them, and that struck them as unusual, and unusual in a medical setting is frightening. Brad joked about the pound of flesh they took, Emily tried to discuss weekend plans, and finally the dermatologist returned. They'd still need more tests, she said, but it was melanoma and he'd have to come back next week. The dermatologist was quiet for a moment. She looked from Brad to Emily and back. "You need to be real nice to each other this weekend," she said, "OK?" It was hard not to read into that.

They tried to make sense of it on the freeway home. Brad had grown up in the sunscreen-optional '70s and '80s, a fair-haired SoCal kid born to burn. That was how skin cancer worked, right? Cancer in general, the consequence of your history catching up to you. Brad had his share of UV rays, but on the bottom of his foot? No sunburn there. The only thing they could relate it to was Bob Marley, who'd had melanoma on his toe. It hadn't worked out well for Bob, but then, Bob had ignored all medical advice. *Be nice to each other*, Brad's doctor had said. They decided not to ignore that advice, either.

At thirty-one years old Brad still held a sense of invincibility. He was an optimist by nature and enjoying a life which seemed to confirm that outlook. He had a big new job at a booming tech startup during a booming tech time and a healthy one-year-old daughter. They weren't rich, but they were confident that they were always going to be OK. It was the new millennium, when hype over an imagined Y2K crash had been steamrolled by waves of new products, services, and technologies rolling out of Silicon Valley. In the San Francisco area it was practically religious doctrine that hard work and smarter tech could hack a solution to anything— even this *thing* on his foot, this *melanoma*. Brad would beat it— *Hell, cut my whole foot off*, he had told the doctor, *whatever it takes*.

But when the tests came back, they showed that taking his foot wouldn't do it. The melanoma had already spread. It had traveled up the leg, as far as the lymph nodes behind Brad's knee. And this was good news, he was told, in a relative sense—it was *just* in the leg, and *only* below the knee. Melanoma wasn't just about skin, though that was where it started. It traveled fast, and when it reached critical organs, especially the lungs or the brain, it was deadly—stage 4. They were calling Brad's stage 3b.

The shock was followed by bravado. They'd nail this thing. The surgeon would take out everything he could see; afterward they'd hit the area with radiation, to kill what they could not. That was the standard of care. Brad wanted something more. Radiation was pre-digital, more 1902 than 2002. Brad wanted cutting-edge extreme. There was something. It wasn't totally new, but it had been approved by the FDA only that year, and "extreme" might have described it. It didn't help most people and the effects weren't predictable. Most people weren't sure that it was really a valid approach to fighting cancer. The drug was an "immunotherapy." Brad figured there was no harm in at least trying.

~

Interferon[1] was one of the most overhyped wonder drugs in memory, but it was also a genuinely important and powerful cytokine. As Brad's oncologist explained, it had helped a small but real subset of cancer patients, especially when used in combination with radiation and chemotherapy. Patient results weren't reliable, and the drug was toxic—in clinical studies, most patients described feeling like they'd had a bad flu for a full year—but the benefits had been sufficient enough to get FDA approval, and the concept of an immunotherapy appealed on a purely gut level. He beat colds with orange juice and sunshine, and he harbored an irrational—though accurate—conviction that his immune system was especially strong, with a *superior* capacity to trump disease.

The protocol was for a year of interferon, most of it self-injected at home. Brad combated the wooziness and flu-like symptoms with manly good humor until, little by little, he started to go insane.

Emily noticed it as a growing irritability in Brad. It was really unlike him to be moody like that, but then again, he had just taken on both cancer and a new high-pressure job—who wouldn't act a little weird in his situation? Then "weird" turned a corner. Brad's conversation increasingly centered on stories about Brad's new coworkers conspiring against him.

He was convinced he had committed the infamous murder of a congressional intern in Washington, DC, that summer. He was behind the pedophile priest scandal too. If it was bad and on the evening news, Brad was responsible. He paced the rooms of their house at night, following voices. One night he urgently waved his wife into the bathroom, then locked the door behind them.

"Be very careful," Brad whispered. Brad told Emily she should call the police. Instead, she called an ambulance. Brad was checked into a psychiatric hospital and placed on antipsychotic drugs. He was also taken off the interferon.

These were difficult weeks for Brad, and maybe even harder on Emily, but gradually, as the antipsychotics kicked in and the interferon left Brad's system, she started to glimpse the man she knew and loved. The psychosis had been a rare and terrifying side effect of the interferon, and it wasn't permanent. That was the good news. The bad news was his planned cancer treatment could not be continued.

His doctor recommended an alternative, a new experimental cancer immunotherapy drug about to start clinical trials. It might be just the sort of adjuvant Brad was looking for—cutting-edge, the newest of the new. One of the study centers happened to be the clinic of his colleague at the Stanford Cancer Center, Dr. Daniel S. Chen.[2]

Dan Chen was a physician and scientist, an MD-PhD with one foot in the oncology clinic, treating melanoma patients, and the other in the immunology lab. As a first-generation child of immigrant academic scientists,[3] his assumed path involved science training and an academic future. Luckily that was where his talent and interest lay, too. He went to MIT to study molecular biology, and, at the last moment, moved to the joint MD-PhD degree track. He moved back west, met a girl over a med school cadaver, and married. Deb, his wife, continued to study obstetrics while Chen completed his medical oncology training and PhD in microbiology and immunology at USC before moving to a postdoc position in Mark Davis's prestigious lab at Stanford, where the T cell receptor was first decoded.[4]

Chen quickly discovered that working both sides of medicine "from bench to bedside" was a rewarding if busy double duty. At the Stanford Cancer Center, Chen ran the metastatic melanoma clinic and saw patients, but he spent much of his time in the lab helping to better understand how cancer and the immune system interacted, and using technology to improve that relationship. Along the way he'd collaborated on a patented chip to visualize T cell interaction with different antigens, a sort of videoscope that showed the signatures of immune response in glowing auras and cytokine starbursts.

Bridging the worlds of cancer clinic and lab—dividing time between the disease as a puzzle, and the disease as life and death—might sound like a natural combination, and it is for some scientists. But pure research and the physician's humanistic art are as different as the jobs of distiller and bartender. In the laboratory it's about the disease, and cancer is both villain and hero. It is insistent of its own existence, and it acts upon the world with a seeming confidence and creativity. And, unlike most normal body cells, it's timeless, a mutated version of self that resists the call to conform and die for the greater good. Meanwhile, the greater good is sitting in the waiting room. In there, in your friend or your patient or your mother, cancer is something else. Something you want to go back into the lab and learn to destroy.

At the time, the great hope for a working cancer immunotherapy was in the development of a cancer vaccine. They'd worked well in mice, and unlike interferon, the vaccines were highly targeted and cancer-specific, meaning few side effects. Dan Chen helped investigate one of the more promising cancer vaccine candidates, a melanoma-specific therapy called E4697 peptide created by Dr. David Lawson. Brad would be among the first to try it—if he was willing to be a guinea pig. Brad didn't mind; as far as he was concerned, the only risk lay in sitting back and simply hoping the cancer didn't return.

Brad liked Dan Chen. He identified with him, and the feeling was mutual. Like Brad, Chen was an athletic California Gen Xer, an ambitious and respected professional who liked good whiskey and played electric guitar and believed he could help reinvent the future. And Chen's vision of that future was extremely appealing to Brad. Chen not only believed that the immune system could be harnessed to help beat cancer, he was enthusiastic about it, and oddly eloquent and patient, even when translating complex immune details that most oncologists couldn't decipher into a story patients could follow. Brad could ask him straight questions and get straight answers back. The two quickly became good friends.

Once a week, Brad would drive across the bridge and sit with Chen. Whenever possible, Brad would take the latest slot in Chen's schedule, so that in addition to the treatment, they could spend a few hours talking through the science. And each week, Chen would glove up and inject 1 cc of the experimental vaccine directly under the skin of Brad's backside. When Brad returned the next week, Dan would palpate the puckered injection site with his thumb. Increasingly, the injection site was like a crater, ulcerous and recessed. It was as if the immune system had responded so aggressively that it had cleared out the tissue in a sort of feeding frenzy. Dan characterized Brad's responses to the vaccine as "amazing, the most powerful immune response I had ever seen."

"Check it out," Brad would say, hitching down the waistband of his shorts. He was proud of his immune system "kicking cancer's ass," and his blood work seemed to back up the physical observations. E4697 had definitely woken up Brad's immune response. But would it help it target and kill his cancer cells?

"You could see that the T cells in the area were doing what you wanted them to do," Chen says. At the injection site, Brad seemed to be rapidly developing what immunologists refer to as "tumor-specific" immunity: a T cell army, geared specifically to a melanoma antigen known as GP 100, something Brad's melanoma cells specifically expressed. The images backed up the blood work. It was too early to say, and he would never say this to a patient, but if what Chen was seeing with Brad continued to bear out, the implications for the field of immunotherapy could be tremendous. For Brad, they would be the difference between life and death.

"But sometimes you see what you want to see," Dan Chen says. "In the case of Brad we saw a strong immune response in reaction to the vaccine." Chen could see it in his antigen visualization machine, cancer antigen–specific responses like signature fireworks. But Brad was an outlier, and his overwhelming local response to the vaccine was far from universal.

Chen describes how the immune systems of the general population react to foreign antigens in terms of a bell curve. Most people live in the middle of the curve, the statistically "normal" response of the immune system. There is a lesser but still sizable population with more extreme immune responses to the right, and very muted immune responses on the left. Brad was on the far right side of the distribution. He responded extremely to almost everything, at least initially. "That seemed good for Brad," Chen says. "But in a way that gave false hope to the field, because he wasn't a typical patient."

For most in Brad's trial group, the vaccine did little to nothing. That was frustrating and heartbreaking. What was billed as a trial felt more like a lottery.

"People see cancer immunotherapy as this huge success right now," Chen says. "It really has been a breakthrough. But the truth is that this success was built on a far longer history of failures. And those failures are borne by the patients."

All data, properly collected, is valuable; even failed studies teach something. The vaccine wouldn't become a viable immunotherapy; in retrospect Chen wonders if perhaps the vaccine did more harm than good. But from that data collection standpoint, Chen's trial was a success. And at the time, it seemed like maybe it had helped Brad, too.

~

Nearly three years after the vaccine trial, Brad was still cancer-free. He hoped his cancer story would soon be behind him, just material for a character-building anecdote in a business speech or political message. He had beaten cancer—*and his tech startup had been bought out by a multinational.* In 2005, when the couples got together for a celebration commemorating Brad's anniversary with no evidence of disease (NED), Chen brought some good bottles of Sonoma wine, and tried to ignore the idea that they were tempting fate. It seemed possible that they'd beaten this thing, or that Brad had, and that the vaccine had helped.

That year the European Society for Medical Oncology conference was being held in Barcelona, Spain. Chen's wife, Deb, had arranged time away from her clinical practice to join him for a week away, just them, without the kids. They were on Las Ramblas on their way to a Catalan meal when Dan's cell rang. The number was Brad's, and he knew right away the news wouldn't be good. He realized then that he had been waiting for this call, bracing for it. Brad's cancer was back, as a new cancer—one that had mutated, upgraded, and escaped.

"There are few things harder than getting a diagnosis of cancer," Chen explained carefully. "But one of those things is thinking that you've beaten cancer, and getting the diagnosis again." For a melanoma specialist, this is an all too familar patern. For a melanoma patient, it is a devastating surprise.

Brad's new cancer had settled along the main artery running through his pelvis and along his iliac crest. Surgery would come first, a deeper cut. After the surgery Brad couldn't feel some of his foot—the procedure had nicked his sciatic nerve—but the good news was that the surgeon had found only one cancerous lymph node among the sixteen examined. It was a hard egg-sized mass, black with dead tissue. The surgeon thought that might be evidence of at least a partially successful immune response. Maybe that was due to the vaccine—it's certainly possible, given that Brad did have years of remission. Chen couldn't be sure. Regardless, the success had obviously been incomplete.

Once again, Brad was going to need a follow-up therapy to try and kill off any cancer the surgery had missed. He asked Dan about the vaccine from their E4697 drug trial. It had worked before, right? So couldn't they try it again?

But Chen knew that, unfortunately, it wasn't quite that simple. Like the immune system, cancer is alive and adapts. Chen's experimental vaccine—all vaccines—did not. It couldn't account for unforeseen mutations, or mutations of those mutations. It's this constant capacity for evasive evolution—"escape"—that makes cancer such an elusive target.

T cells activated by the vaccine might have killed the rogue cancer cells that expressed that antigen. But the vaccination had been too local; it had not globalized through Brad's body. And it hadn't been re-upped afterward—it wasn't possible or ethical. The trial had failed and closed. The surviving cancer cells had remained, invisible to scans, growing and mutating.

In a perfect world, Brad could get a new, better vaccine to match the new antigens in his new mutated cancer, the way we get new flu vaccines every year to match the latest influenza strain. Making such a vaccine would require quickly sequencing the entire genomes of both the patient and the cancer; it would require powerful bioinformatics algorithms, run by computers that did not then exist, to compare all the proteins of Brad's body cells against those of his tumor cells, and recognize the best unique cancer antigens for his T cells to target; finally it would require the technical ability to quickly translate all that data into a personal vaccine.[5] We can do that now; in 2006, that perfect world was science fiction.[6]

~

Brad was still recovering from the surgery, so Dan phoned Emily with a suggestion: Brad should try to get into one of the clinical trials for a promising new form of cancer immunotherapy called checkpoint inhibitors. Chen had been excited about the possibilities of Allison's discovery for years, and now experimental trials were about to begin, run by the two drug companies making competing versions of the anti-CTLA-4 antibody. One was called tremelimumab, manufactured by Pfizer; the other was Jim Allison's version, ipilimumab, made by Bristol-Myers Squibb. Dan's colleague at the University of Southern California, Dr. Jeffrey S. Weber,[7] was in charge of one of only three such trials, testing for safety.

Chen had witnessed the ferocity of Brad's immune response firsthand, a good indicator that he might respond well to the new immunotherapy. Brad wanted in. Could Dan help?

Chen could recommend him, but ultimately, it was not his call. He knew Dr. Weber, as they had worked together studying vaccines and cytokines. He also knew that Weber's team was deluged with calls from the physicians of desperate patients around the world. The buzz around the anti-CTLA-4 trials was electric. *Everyone* wanted in.

Dr. Weber had a reputation as a caring, thorough clinician. His requirements for patients to be considered for this drug trial were appropriately rigorous.

Dan wrote Weber, giving him Brad's medical history and numbers and telling him more anecdotally that Brad MacMillin was the greatest immune responder Chen had ever personally witnessed. Chen couldn't push it, but he made his case. Weber wrote back: Did he want to refer the patient?

Brad started on the anti-CTLA-4 trial that fall.

~

In some patients, blocking the CTLA-4 brakes might mean the difference between having a T cell response to cancer and not. For patients like Brad, who had a hair-trigger immune system already teetering on the edge of autoimmunity, taking off the brakes would turn out to be a very dangerous ride.

"Brad had just a crazy response to it," Chen remembers. What the anti-CTLA-4 drug unleashed in Brad's immune system was something more like a riot than a precise military operation. Brad got his first injection of the experimental MDX-010 antibody on October 5. Within a week he had an extensive rash on his neck, arms, and face, and a huge welt on his thigh near the injection sites. And every day it got worse.

"Brad was really, really sick," Dan says. "He couldn't eat for over a month and ultimately had to get slammed with some of the strongest drugs we have to shut off the immune response." Brad checked into the hospital under Dr. Weber's care the day after Christmas. He had already lost forty-five pounds and endured weeks of misery. The experience of having his own immune system attack his guts was, Brad later said, the most brutal thing he had ever been through.

An examination would show that the extreme immune response had decimated his GI tract. Had it been enough to completely wipe out Brad's cancer, too? Only time would tell.

Slowly, Brad bounced back from the anti-CTLA-4 trials. He was clear of cancer for 2007, gaining weight back, feeling his old self again into the new year. The family Christmas letter was hopeful, if cautious, and the following August he wrote to tell his friend that the PET/CT scans and brain MRI were all still clear. "This marks my 2-year NED anniversary again," Brad noted, but he didn't think they should have a celebration. "Don't want to tempt fate this time!"

And anyway, he knew Dan was especially busy now, with three kids, a busy oncology practice, and a new job in biotech.

∿

In 2006 Dan Chen had accepted a position at Genentech, whose steel-and-glass laboratories fronted the San Francisco Bay. Its open floors and dedicated buildings were cutting-edge. This was a place filled with academics, but it wasn't academia. It was a powerhouse of resources for new drug development.

Taking care of patients was still hugely important to Chen,[8] and he'd kept his clinical position at the Stanford cancer center. And Brad was still one of his patients, or really a former patient turned friend, a guy who could remain free of cancer. Brad had been getting regular scans since his last treatment and they'd been clear, and finally by the end of 2008 he and Emily were confident enough to stop scanning the horizon for smoke and start making plans for a fuller future. That next fall Chen got an email from Brad and Emily giddily announcing the birth of a daughter. Five months later Brad wrote again. He'd had his scans. The melanoma was back. In the same place, on the inside of his hip.

Dan peppered his friend with questions and options—had he contacted his surgeon yet? Had they tested the tumor for specific mutations? Was Brad considering one of the cytokines that had been approved, like interleukin-2? It wasn't perfect, Chen said, it was a general immune treatment and there was always a risk with a hair-trigger immune responder like Brad, but some patients had responded favorably. And most important, Chen said, Brad hadn't tried it yet.[9]

Brad was running out of options. There just weren't that many pages left in the oncologist's playbook in February 2010, not against melanoma. Silently, Brad and Emily stopped thinking in terms of beating this thing. The goal now was to keep it in check, to keep it from spreading again after the next surgery. That would be enough of a win. Brad and Chen went back and forth for a few months while Brad shopped his options and tried to decode the clinical trials he found on the internet. Finally he settled on a targeted therapy called Gleevec. It wasn't an immunotherapy; it had nothing to do with the immune system. Gleevec was a small-molecule drug, taken by mouth, that interfered with the metabolism of some types of cancer. In 2008 the drug had been compared to a "magic bullet" and "a miracle drug" by a few hopeful medical journals. Others called it "a breakthrough in cancer treatment."[10] That sounded good. The drug had shown great results in patients with a specific genetic mutation[11] that caused a form of leukemia. Brad had neither the mutation nor leukemia, but there was hope the magic bullet might help other cancers, too. It was worth a shot. He could get his primary oncologist at the University of California, San Francisco, to prescribe it "off study" and, remarkably, his insurance would pay.

"I'd recommend you try it with an immunotherapy," Chen said, maybe IL-2. If they were going to beat this new cancer, Chen said, it should be right now, immediately after surgery. But Brad had already suffered enough side effects, and IL-2 was famously rough. He'd told Chen he'd "keep it in his pocket, in case," and stick with the Gleevec. That strategy worked until it didn't, and in the spring of 2012, Brad got the news. He was stage 4. The melanoma had metastasized to his liver, perhaps elsewhere nearby. He knew it was a bad diagnosis, but he still hoped to attack it, for the fifth time now, as aggressively as possible, and he had a pretty good idea how.

Cancer science had made important progress over the eleven years that Brad been a cancer patient, either in remission or treatment therapy. The experimental checkpoint inhibitor drug Brad had tried back in 2004 was now an FDA-approved therapy called ipilimumab. Blocking the CTLA-4 brake on T cells was transformative for some cancer patients, but had proved too much for Brad's hair-trigger immune system, so that drug was ruled out as a viable option. But ever since he and Dan Chen had been friends, Chen had been excited about another discovery, a second checkpoint. And in recent months, that excitement had skyrocketed. Now Brad hoped that Chen could leverage some of the new developments in his life to help Brad save his.

~

Genentech didn't have an immunotherapy drug pipeline when Chen arrived in 2006, and it was about this time that Chen realized that his boss in charge of the patient side of early drug development, company VP Stuart Lutzker, MD, PhD, was a cancer biologist. In fact, most of the cancer people at Genentech were cancer biologists. "And cancer biologists hated immunotherapy." He laughs. "I Mean, they hated it!" Much of the history of the field had given them cause to. But for whatever reason, the company had hired a number of cancer immunotherapists.[12]

One of those was Ira Mellman. Mellman had a distinguished twenty-plus-year career that included postdoc work in the lab of Ralph Steinman, the eminent New York–based Canadian physician and medical researcher who'd discovered the dendritic cell (and in 2011, received the only Nobel Prize ever awarded posthumously).[13] Mellman himself had been a department chair at the Yale School of Medicine and the scientific director at Yale Cancer Center, and his name was in the back of every cell biology book. He had tossed all that off to come make molecules at Genentech.

There was an upside obviously, but for Mellman the decision had less to do with career or money than with family and friends—both of his children suffered chronic inflammatory disease, and each year saw more friends succumbing to the ravages of cancer. "To see that, and then to be presented with the opportunity of moving to the best place on earth to do drug discovery—I don't know if it's a moral obligation to act on that," Mellman explained. "But it certainly was a motivating force for me."

The Genentech brass met twice a week in what served as essentially the pilothouse for a massive corporate ship. Mellman's boss on the new molecule-making side was Dr. Richard Scheller, a Lasker award–winning biochemist and VP in charge of Genentech's Research and Early Development Organization. Ultimately, charting the next course was his decision, and in the room around him each week there was a lot of discussion about exactly what that course should be, characterized by Mellman's "secret" immunotherapists on one side and the cancer biologists on the other. While nobody wants to say they were heated, they were definitely "lively,"[14] and they got livelier with the developments around a molecule called PD-1. If CTLA-4 had cracked open the possibilities of cancer immunotherapy, PD-1 was threatening to blast it wide open. At least, that was how the immunologists saw it.

∿

Like most big discoveries, PD-1 was found by researchers looking for something else. In this case the something else was the body's natural quality control mechanism that weeded out dangerous T cells before they entered the bloodstream.

As immune researchers knew, T cells emerged from the thymus. Each had a different antigen receptor that had been randomly assigned, a lottery-ticket approach to preparedness against unknown antigens.

T cells triggered only by foreign non-self antigens were good defense. But T cells that happened to be randomly assigned with receptors activated by the body itself—T cells that happened to have lottery tickets for self antigens—were dangerous. If they went out the door they would attack the body, resulting in autoimmune diseases such as lupus and multiple sclerosis. So, in a tidy bit of immune housekeeping, these T cells were instructed to self-destruct.

Scientists called the T cell's self-destruct signal "programmed death," or PD. PD is built into every T cell, just in case. That receptor was activated by a ligand—a key that matched it, bound to it, and activated it. But so far nobody had actually located the programmed death receptor, or the ligand.

In the early 1990s an immunologist named Tasuku Honjo and his colleagues at Japan's Kyoto University were trying to find the genes responsible for PD, as a way of identifying the PD receptor. Honjo devised a process of elimination;[15] what was left he assumed to be the genes he was looking for. Honjo called it "Programmed Death-1," PD-1 for short.[16]

He was wrong about what he'd identified—it wasn't the self-destruct signal (but it would keep the name). In fact, they didn't know what receptor the gene corresponded to or what it did, but mice that lacked the gene gradually showed signs of a lupus-like disease. Honjo believed they'd found an important aspect of controlling autoimmune disease, and continued his work.

This is where the story gets complicated, or at least litigious, since the subjects in this story are currently wrapped up in lawsuits. Not every discovery is immediately understood for its context or full significance—many, in fact, are not. And sometimes researchers find puzzle pieces that are only known to be missing because another researcher found the corresponding piece, and didn't know it—that happens frequently as well. Also, not every immune discovery was viewed specifically in the context of the immune system's complicated relationship with cancer. As a result, assigning absolute credit for a collective lightbulb moment isn't particularly useful. What's important is that several researchers around the world were using new gene sequencing and imaging technology and asking questions about genes and cell receptors, the immune system and cancer. And several of them, independently or together, found pieces of the PD-1 puzzle. Honjo definitely found what he found; whether he knew more than that, and when he knew it, will be decided by history. Meanwhile Honjo wasn't the only one looking for the other side of the PD-1 receptor on T cells. Those others included doctors Gordon Freeman, PhD, and Arlene Sharpe, MD, PhD, a bespectacled husband-and-wife team at Harvard, and Dr. Lieping Chen, a Beijing-trained oncologist with a PhD in immunology from Drexel University and a lab at the Mayo Clinic.[17] Each of them recognized pieces of this particular immune puzzle and contributed to the final understanding of what PD-1 was and what it could do.

Lieping Chen had watched the attempts at making cancer vaccines. He had seen the glimmers of success from what Dr. Steve Rosenberg's team at the National Cancer Institute and others had accomplished in boosting immune response to cancer by boosting the T cells, and he had also seen the limitations of that approach. These approaches undoubtedly made powerful T cells, improved in quantity and quality. Cancer vaccines generated those additional T cells inside the body, while cellular approaches like those in Phil Greenberg's and Steve Rosenberg's labs identified the T cells from a cancer patient's blood that recognized the right cancer antigen, fertilized them into a ninety-billion-strong army, then injected that army back into the bloodstream. So if these approaches worked at improving the T cells, why did they not reliably work at attacking and killing tumors? It was a paradox that bothered nearly every cancer immunotherapist.

"I'd already committed to cancer [as a career], so I had to remain positive," Lieping Chen explains. "Others would think, 'Oh, it's pointless to do cancer immunology, T cells, no good! Just leave the field!' But those who stayed believed there's something here. It works in the blood, but not in the body. Why? It had to be something in the tumor microenvironment—something in the tumor, working against [the T cell attack]."

Lieping Chen had started to work on investigating that environment in 1997, and in 1999 he reported finding a molecule that was expressed by some body cells but especially highly expressed on certain tumors, and potentially involved in down-regulating (turning off) immune responses.[18] He gave it the name B7-H1. In 2000, Freeman (informed by some work from Honjo) published work identifying this same molecule, highly expressed on some tumors, as the other side of the PD-1 handshake, the ligand of that PD-1 receptor, the yin to its yang. Together that meant the molecule in question, Lieping Chen's B7H1, was also the sought-after Programmed Death number one ligand. Freeman and Sharpe had mapped out both sides of the biology. They called this molecule PD-L1. (These molecules *still* had nothing to do with programmed death, but that was the name that stuck.)

The net result was that two sides of a receptor-ligand handshake had been identified; when the ligand plugged into the receptor on the T cell, the T cell stopped an attack.

Maybe any one of these researchers could have figured out all the pieces on their own, and maybe they did; subsequent scientific prizes have recognized all of them equally as codiscoverers. But regardless of credit and ownership, what matters is that the effort had identified an especially intriguing and important molecule pairing. The PD-1 / PD-L1 interaction seemed to work as a sort of stop sign for the T cell, like a secret handshake given to the T cell up close and personal, telling it not to attack.

This is a useful communication when the seemingly "foreign" cell the T cell is attacking turns out to be a developing fetus. But PD-L1 had been discovered to be prevalently expressed on cancer cells, and for a similar reason. It shut down (or down-regulated) immune response.

Though it had not yet been proven in humans, the belief was that the interaction of PD-1 and PD-L1 told the T cell attacker to stand down. It was a secret handshake between body cells, coopted by cancer cells—especially heavily mutated cells—in order to evade recognition and attack by T cells. PD-L1 made a cancer cell look like a normal body cell. Even to activated T cells, already massed at the tumor, primed to kill, the PD-1 / PD-L1 handshake told the T cell to stop. The race was on to develop antibodies to block that secret handshake, and test them as a possible immunotherapy for cancer.[19]

Sharpe and Freeman at Harvard had published their PD-1 pathway patents first, but they allowed the intellectual property to be distributed nonexclusively. That allowed any lab in the world the right to make an antibody to block it. Spurred by the success of the CTLA-4 checkpoint blocker ipilimumab, drugs to block the PD-1 (T cell) side of the handshake were fast-tracked, with seven pharmaceutical companies licensed to produce the antibody. In 2006, humanized anti-PD-1 antibody was finally produced in quantities sufficient to begin clinical trial as a cancer medicine.[20] The race to develop and test an antibody to block the PD-L1 (cancer cell) side of the handshake followed quickly behind.

~

On December 10, 2010, Genentech employees were still wrestling over whether the company should jump into the immunotherapy race and create an antibody that blocked PD-L1. The project would represent a daunting gamble;[21] at this point no checkpoint inhibitor had been approved. One of the anti-CTLA-4 drugs, which had started testing nine years earlier, was still burning cash in its protracted clinical trial reboot; the other had been abandoned by its company during phase 2, a wreck by the roadside cautioning would-be travelers. For Dan Chen on the clinical-trials side, Mellman on the molecule-development side, and the rest of the underground cancer immunology hopefuls in the conference room at Genentech, it felt like now, or probably never. "We were at least able to argue our case that this was something new and we at least needed to try it," Chen says. "Even if nobody was listening, we felt it was important to present an argument that this was very different[22] than the other approaches we had to treat cancer, and offered a very different value proposition for patients."

Mellman talked the board through the scientific arguments and the new data. He believed that even "the crummy mouse models" seemed to demonstrate a mechanism of interaction between cancer and the immune system. "That meant, whatever the arguments, in the aggregate the data (on PD-1 / PD-L1) was strong enough to act upon." Chen added what he had seen in his own clinic and his own patients, like Brad. If there was a possible upside that might be measured in more than weeks or months, one that was fundamentally different from what other therapies were offering, even if it didn't work for everyone, the cancer patients in his clinic wanted to at least have a shot at those kinds of durable, transformative responses.

The debate raged on for hours, Chen remembers, until finally, Scheller said, "Enough, this is ridiculous—we're moving forward." When Scheller pivoted, Chen says, the room pivoted, and the company behind them. Mellman was amazed. The new PD-L1 team was tasked to show progress in just six months; if it didn't work, they could scrap it with minimal fallout.

The time line was impossibly tight but was helped by a lucky break languishing in the Genentech laboratory. Years earlier, Genentech researchers had also come across the PD-L1 ligand, and patented an antibody that targeted it. At the time the ligand had been just another protein on tumor cells to be numbered and catalogued,[23] a potential bull's-eye for the usual sort of cancer drugs pharmaceutical companies specialized in, the ones adding months to patient's lives. Now it would provide a leg up on bringing a radically new type of medicine to the clinic.

The anti-CTLA-4 drug was still being tested in blinded clinical trials, so the results weren't available. But it was clear that CTLA-4 was a checkpoint that prevented what immunologists call the T cell's priming and activation stage. PD-L1 seemed to be involved in a different type of T cell inhibition. It wasn't the activation phase. PD-1 / PD-L1 seemed to shut down T cell attack, long after the T cell had been activated. Possibly, this explained what cancer immunologists had been seeing under the microscope. They saw T cells that had been primed by the battle cry, activated by tumor cell antigens, cloned up into a billions-strong T cell army, and marched, massing at the frontier of an antigen-presenting tumor. They had the "go, kill" signal. They were ready to attack. But then, for some reason, nothing happened. The T cells stopped. They did not attack the tumor.

Was PD-1 / PD-L1 the secret handshake, given up close and personal at the T cell / tumor battle line, that explained this strange failure of activated human T cells to kill those cancer cells? This checkpoint inhibitor certainly matched Chen's hypothesis of what was happening in his clinic; plenty of other cancer immunologists around the world thought so, too. PD-1 / PD-L1 fit the observations. It looked exactly like the missing piece of the immunity puzzle. And it had yet to be tested.

~

This was a particularly thrilling time for Chen and Mellman. They were cancer immunologists who actually got to make a cancer immunotherapy drug they believed would work, and they knew how fortunate they were. They had a green light and funding, teams of top researchers, and a potential PD-L1-blocking antibody already on the shelf at Genentech. Their task was to turn that into a real drug for real patients. It wasn't easy, but for once, it seemed possible.

They started with mouse models. The PD-L1-blocking antibody worked there; it seemed to reopen the road of a stalled immune response to the tumor by blocking the tumor side of the PD-1 / PD-L1 handshake. Once again, cancer was cured in mice. The next step was to make human antibodies against PD-L1 and see how they did blocking the tumor's handshake in people. Chen would be in charge of the trials.

Six weeks later, in February of 2012, his team got the first scans from their phase 1 clinical trial. The first responder was patient 101006 JDS: Jeff Schwartz. That was a thrilling moment, but it was just one patient, with kidney cancer. Chen's boss on the clinical trials side was Dr. Stuart Lutzker, who'd told Chen he'd believe this immunology approach when Chen could show him proof it worked in lung cancer, the leading cause of death globally and Lutzker's specialty. But Chen had scans from a lung cancer patient too. He wasn't a full responder yet, but *something* was changing after he went on the drug. Chen saw that the patient's tumors, which previously had a rounded mass, now had a spiny appearance, as if the tumor was pulling back and retracting along these spikes rather than continuing to grow into the surrounding lung. "Each tumor type has a sort of personality," he explains, "a signature unique to it, and when it begins retreating and dying, that also has a unique appearance on the scan." Chen remembered bringing the scans up to Lutzker's office. "He just looked at it and said, 'This isn't what a growing tumor normally does. This is real.'"

"It just transformed his opinion one hundred eighty degrees. And remember, he's a cancer biologist," Chen says. His boss had agreed to the company direction, but until then, Chen felt he wasn't fully convinced; he knew the history of immunotherapy. There was a chance that this, too, would end up being one of those stories that didn't translate reliably into a drug for a human cancer population, an expensive and humiliating dry run. "But right then he went from being very much against it, to OK'ing the whole new direction."

Until it was made public, Chen couldn't have access to all the data from the anti-PD-1 drug trials, which were further along but run by another drug company. But for anti-PD-L1, he was the center of the data spiderweb, in touch with all the clinical investigators. "We immediately started seeing responses," he says. "And these responses were unlike anything we were used to. They could be sudden, they could be transformative, they appeared to be durable, and they were happening in patient types we don't usually associate with being responsive to an immune-based therapy, like lung cancer. And some of these patients were reporting their tumors shrinking within days or a week!"[24]

What's more, the PD checkpoint was turning out to be far more specific than CTLA-4. Releasing the brakes on the T cells by blocking CTLA-4 resulted in a body full of T cells without brakes, and some serious toxicity from an immune army suddenly unleashed. It would also later be discovered that blocking CTLA-4 resulted in a diminished number of immune regulatory cells throughout the body,[25] resulting in a more generalized immune response, and greater toxic side effects. But the PD checkpoint was activated only right at the moment of tumor killing. Blocking that checkpoint had fewer toxic side effects and, for those patients that responded, produced some dramatic results.

This was what Chen was working on when he and Brad met for lunch near Dan's offices in January 2012. It was a catch-up session, but also an informal medical consultation for Brad. "Cancer patients like me need more doctors and scientists like you, working for a cure," he'd told Chen. Brad was especially enthusiastic about the PD-L1 thing Dan was working on. He was still disease-free, but he was realistic, too.

Four months later, Brad learned he was stage 4. The cancer was in his liver, and he was looking to Dan Chen for new options. Maybe, he wondered, a "Hail Mary" with IL-2? Or perhaps the new experimental drug Chen was so excited about. "I think you called it anti-PD-L1?"

Chen carefully ran Brad through his options. IL-2 wasn't the "Hail Mary" option now. It hadn't proved to be so effective for liver metastases in general, though there was a chance—Brad was such a strong responder, maybe it would work for him. Meanwhile, yes, if Brad was interested, there were several trials open for both PD-1 and PD-L1. "You'll have to see if you're a candidate, but there are many—including one with Dr. Jeff Weber, now in Tampa," Chen told him. Brad wanted something closer. Chen said Weber's PD-1 trial would be worth the plane rides. "If you qualify," Chen said, "I'd go for that."

Brad found a PD-1 study, then dropped off the radar. When Chen next heard from him, Brad seemed despondent. Nothing was working. Next he tried IL-2 but he saw no improvement. It had been a long twelve years.

Radiation, chemo, vaccines, two different cytokines, and the newest checkpoint inhibitors—by 2013, Brad had practically lived the modern history of cancer treatment. And he'd beaten the odds. But he hadn't beaten cancer.

Now Brad was tired. He'd tried, he and Emily both. Chen asked if Brad wanted his input, told him he was thinking about him. Brad didn't like the tone of that. Yes, he said, he wanted help—didn't Dan remember the last time they'd talked about his treatment options? Over lunch, Chen had talked at length about the virtues of his new experimental drug. But now Brad wasn't eligible for Chen's PD-L1 studies, not after all his previous immunotherapy treatments. If Dan could use his influence to get him the drug, or had any new ideas for him at this point, he'd be grateful. Otherwise, what was the point of talking?

Chen and Brad had long before crossed the line of the doctor-patient relationship. This was personal. And now Brad was taking Chen's failure to cure his cancer personally. The relationship between physician and patient is an intense journey, often years in the traveling. Sometimes that intensity can be a liability.[26]

A few months later, Brad wrote again. He had decided to enter a study at MD Anderson Cancer Center in Houston, something with tumor-infiltrating lymphocytes. It wasn't what Chen had recommended, but Brad had decided this was his best shot. "Thanks for your input," he wrote. It was the last Dan Chen would hear from his friend.

∾

Emily can't speak for Brad, and Brad can't speak for himself. But Emily has no regrets about the therapies they tried. Nor does she feel resentment about the immune therapies that didn't ultimately work for her husband. *Regret* and *resentment* are the wrong words. At that time, it felt like a foot race between one man's disease and the pace of global cancer research.

"We had always said, it's only when the doctors tell us they have no more options that we'll get depressed," Emily says. Finally, they did run out of options. But both Brad and Emily felt it had been a good race. It was a story, and one she wanted to have shared and written here, partly to preserve the memory, partly to thank Dan for his friendship. But mostly so that others might learn from it—especially patients whose futures are still to be written.

Could it have gone differently? Would it now? Maybe a different vaccine, used for a longer time or re-upped, would have helped him; maybe in combination with a checkpoint inhibitor it would have cured him. A million maybes, and not enough time.

The year 2014 doesn't seem so long ago, but in terms of immunotherapy, it's a lifetime. Certainly it was Brad's. Oncologists now tell their patients that the goal isn't necessarily beating cancer today; it's staying alive long enough to take advantage of the next advances, the ones right around the corner. But in the end, the science didn't catch up with Brad. Cancer immunotherapy was a breakthrough as a proof of concept; Brad needed a medicine. It's a caution about hype and hope. The breakthrough is a door, now open; the beginning, but not yet the cure.

~

Any oncologist who dwells too much on the patients who don't make it won't last long in the field, and when Chen started his clinical practice that was especially true for melanoma, where survival rates guttered in the single digits. The breakthrough changed that outcome for those patients, and many others. It changed their options.

Patient 101006 JDS, Jeff Schwartz, was his first complete responder, the first patient whose scans Chen looked at and realized, *It's gone.* Chen worked in a science built on stories. Most were bittersweet, like Brad's. Jeff Schwartz was different. He was the first patient Chen had personally witnessed beating cancer with his immune system. It was as close as he'd come to a Coley moment, what Steve Rosenberg had seen in his VA operating room back in 1968, or Jedd Wolchok had witnessed at Memorial Sloan Kettering as a teenager.

"I'll never forget Jeff," Chen says. "I had almost rejected him from the trial, he was so sick. And a month later, I get back this email from the doctor that was treating this patient. I cried when I read the email. This patient that could barely get out of bed before he started treatment on our experimental trial, only four weeks later, is going to the gym three times a week. And—this drug had given him his life back."

Here, finally, was the upside, emotionally, of splitting his time between the clinic and the lab.

"We don't see this sort of thing often in our careers. Or in our lifetimes," Chen says. "And to see it and be at the heart of it—I cannot tell you how exciting and rewarding it was. This is what we always thought might be there, but nobody believed was really there. And the truth is, it actually worked better than we had ever hoped for. We'd always had a vision of what success was going to look like—and this works faster than we had ever dreamed of. We thought, to get the kind of responses we're seeing, it was going to take a cocktail of drugs—because that's how complicated the biology is. And so this is a case where the clinical experience—when you see something that's unexpected, you go back and you learn from it.

"We're at a breakthrough point in our battle against cancer," Chen says. "This is our generation's moon shot. And it's just the beginning. Think of how far antibiotics came after the discovery of penicillin. That's been decades. We've only just discovered the checkpoint inhibitors—PD-1 only saw its first approval in 2014. So it's the breakthrough; we've just discovered our penicillin. But it's only the beginning."

Chapter Seven

The Chimera

Cancer immunotherapists had spent decades trying to find just the right T cell among the hundreds of millions in the bloodstream, one that could recognize the specific antigens on a patient's tumor. Then they'd spent more time patiently trying to grow these T cells, and get them to attack.

Meanwhile, another group had a different approach: engineering their own Frankenstein T cell, stitched together from various parts in the lab, designed specifically to seek and destroy a patient's own cancer.

This new invention, a monstrous T cell assemblage, is a sort of immune-cell chimera (in Greek mythology a chimera was a patchwork monster with parts from a lion, goat, and serpent), and so it's called a chimeric antigen receptor T cell. "CAR-T" sounds much better.

CAR-T is just a reengineered human T cell. It is often called the "most complex drug ever created,"[1] if only because it's not a molecule or antibody like other drugs, but a whole cell that has been removed from a cancer patient, tweaked in the lab to recognize that patient's cancer, and then injected back into the patient. What sounded like science fiction when research started was approved by the FDA in August 2017 and is now manufactured in New Jersey with a twenty-two-day turnaround.

∽

The engineering is complex, but the concept is simple. T cells[2] hunt and kill only what they're programmed to "see." And the business-end of that "seeing" is the T cell receptor, or TCR.

The hope was that if you change the TCR, you change what that T cell targets. And maybe you could get it to target disease.

And that was exactly what occurred to a charismatic Israeli researcher named Zelig Eshhar. In the early '80s he'd started thinking about how the business end of the TCR, the part that "sees" its matching antigen, worked a lot like an antibody.

Each TCR is stuck into the T cell surface like a protein carrot, but the part that reaches outside the cell and recognizes the shape of an antigen is a lot like the little grabby protein claws of antibodies. Eshhar could imagine popping off the end of the TCR and popping on a new antibody like a vacuum attachment. In fact, you could have an infinite number of attachments, each specific to recognize and bind with a different antigen.

Turning the theory to reality required a fancy bit of bioengineering, but in 1985 Eshhar produced a simple proof of concept.

He called his primitive CAR a "T-body." It was a T cell retooled to recognize a relatively obvious antigen target that he had selected. (It happened to be a protein made by a fungus called *Trichophyton mentagrophytes*, better known as athlete's foot.) This humble experiment belied mind-blowing possibilities.

By 1989 Eshhar had been persuaded to spend a sabbatical at Steve Rosenberg's lab at the National Cancer Institute, where he would work with a number of brilliant young physicians, including Dr. Patrick Hwu. The lab's IL-2 and T cell transfer work had yielded some new findings, and Hwu was trying to use them against a larger group of cancers.

His project involved inserting a gene for tumor necrosis factor (TNF) into the specific subset of T cells that had recognized tumor antigens and nosed their way into tumors. These "tumor-infiltrating lymphocytes," or TILs, were in the perfect position to continue their mission and attack the tumor; instead, for reasons that were at that point undiscovered, they just sat there, their attack called off by tumor tricks like PD-L1 and others within the tumor microenvironment.

Hwu's interest was in turning those TILs into little guided missiles that would dive into a tumor and express their TNF cytokine payload. Those guided missiles needed a customizable guidance system to target the different tumor antigens. "Zelig had shown that an antibody and a T cell could be combined to target something," Hwu explains. "Now the question was, could we get it to target cancer cells?"

Hwu already had a good deal of experience putting new genes into T-cells. "It was really hard to do that in the 1990s," Hwu recalls. Until they worked out a method for using retroviral vectors as delivery vehicles—or, more recently, CRISPR—that work involved sticking a little needle into a T cell and micro injecting one at a time. "Zelig and I spent a lot of time together," Hwu says. "A lot of all-nighters in the lab." The work built on the proof of concept provided by Eshhar's T-body, genetically engineering the T cells to change their TCRs and target something else.[3] It took years to develop and it didn't work well, but it did work, and the resulting paper heralded the new CAR-T name and some enticing possibilities. They had successfully replaced the T cell steering wheel, and by doing so changed where the T cell wanted to go. And most crucially, they'd changed a T cell's target destination to a specific cancer.

Part of what prevented these early CAR-Ts from being an

effective cancer therapy is that they were lemons that got bad mileage; the robo–T cells didn't last long enough to replicate themselves or finish the job of killing cancer. It would be the work of Memorial Sloan Kettering Cancer Center researcher Dr. Michel Sadelain that provided the clever workaround for this and several other engineering problems, creating a truly "living drug." Sadelain also gave his new CAR an important new target—a protein called CD19, found on the surface of certain blood cancer cells. The result was a sleek, stylish, and self-replicating second-generation CAR with plenty of fuel and an important destination.[4] Sadelain's group shared the sequence of their new second-gen CAR with Dr. Rosenberg's group at the National Cancer Institute, as well as with the head of a lab 150 miles north of Bethesda led by University of Pennsylvania researcher and physician Dr. Carl June, who would borrow and build on these ideas and others,[5] and add some of their own.

Three groups were now pushing toward moving this breathtakingly complex and powerful experimental cancer therapy into first-in-human trials. In a sense, their work is inseparable; at times, they had worked together. But for most of the world, it would be the trials June's team was about to undertake that would provide a first introduction to our brave new future of CAR-T.

∽

The technology of placing genes into cells had come a long way since Hwu had started injecting them by hand. The foreman in this modernized CAR assembly line was the repurposed shell of the virus that causes AIDS. Instead of delivering disease, the repurposed virus would "infect" a patient's T cell with new genetic instructions that reprogramed it to make a different sort of TCR, one that targeted only a protein[6] found on the surface of B cells afflicted by the most common form of childhood leukemia, called acute lymphoblastic leukemia (ALL).[7]

The HIV virus is especially well suited for this job, because, like leukemia, AIDS is a disease of the immune system. The human immunodeficiency virus is specialized to target and infect T cells—specifically the body's helper T cells, the ones that choreograph immune response to disease like cytokine-spewing quarterbacks. The virus renders them useless at this job, resulting in a shutdown of adaptive immunity in the body and the disease we know as acquired immune deficiency syndrome, or AIDS.

In the 1990s June was a certified leukemia specialist gaining firsthand experience with the HIV virus's stunningly efficient genetic delivery system at the NIH, where he collaborated on an experimental CAR-like treatment that redirected killer T cells to hunt down infected helper T's in AIDS patients.[8] He also developed techniques for growing out T cells derived from human donors that were robust enough to last decades. The first CAR clinical trial in humans would be for HIV. Early phase data looked good, but before the work was complete it had become unnecessary, due to the development in 1997 of the first protease inhibitors, drugs that blocked the HIV virus from replicating. Overnight, these drugs changed the prognosis for millions of people, as well as the direction of June's career. He now moved his work and practice over to a laboratory at UPenn and the Children's Hospital of Philadelphia, and intensified his focus on another disease, one that had recently become extremely personal.

In 1996 June's wife, Cynthia, had been diagnosed with ovarian cancer. When Cindy June didn't respond to traditional therapies, Dr. June had turned to immunotherapy approaches still in their infancy, and had his lab make a customized version of another lab's immunotherapy vaccine that had shown some promise. It was called GVAX.[9]

"I had no idea how hard it was to turn a laboratory experiment into a clinical trial," June said. He felt GVAX was a therapy ahead of its time, and he believed his wife had a good response. But, as with all cancer vaccines of the era, the effect did not last. June suspected that the tumors were somehow turning that immune response off. "I knew the work of Jim Allison," he remembers. "I knew in mice, his antibody made immunotherapies work better, and so combining them was a no-brainer." June tried repeatedly but was denied access to the precious anti-CTLA-4 antibody from the manufacturer. "It was very frustrating," he says. Cindy June passed away in 2001 at the age of forty-six. June funneled his grief for the mother of their three children into his work and moved his full-time focus on a CAR for cancer "to the front burner."[10]

Nine years later, it was ready. One of the first to try it would be Emily Whitehead, a six-year-old girl with ALL and no options. Eighty-five percent of kids with ALL respond well to traditional therapies; Emily was in the 15 percent who didn't live long.

Emily had already endured twenty months of chemotherapy. The treatments had bought her only an extra few weeks.[11] The cancer was doubling daily in her bloodstream, and a bone marrow transplant was no longer an option. Finally Emily's parents, Tom and Kari, were told that their daughter probably wouldn't last the year. Her oncologist suggested putting the girl in hospice. The horror of that notion made their next decision conceivable.[12] When the UPenn team received approval for human trial in 2010, they had no illusions about the risks involved, or the stakes for their first experimental pediatric patient.

~

Viruses exist at the edge of our definition of life. They're not made of cells, but are essentially just genes with legs in a protein shell.[13] They cannot reproduce on their own, relying instead on the larger, more complex cells that they infect to process their genetic blueprints for them. In the case of the HIV virus it's human T cells that they find and inject with their DNA. HIV is devastatingly effective at infecting T cells. That made it an ideal carrier for a CAR-T's genetic blueprints.

In June's lab the HIV virus was emptied and fitted with new genetic instructions. Then it was introduced to Emily's T cells, which had been carefully separated out of her drawn blood. Now, rather than inserting genes that told the T cell to make more versions of the virus, the injected genetic instructions turned a killer T cell into a programable cellular assassin.

In Emily Whitehead's case, those T cells would be reprogrammed to target the CD19 proteins that marked her own sick B cells. In a healthy human, B cells are essential aspects of the normal immune system; in patients like Emily, those B cells had mutated and become cancerous. (B cells, when centrifuged out in mass, appear white. Science used Greek roots to turn white [leuk] blood cells [cytes] into "leukocytes"; we call cancers of these cells "leukemia.")

Over the previous weeks at the Children's Hospital of Philadelphia, Emily's blood had been drawn and centrifuged and some T cells selected out. Those T cells were then infected with the virus that would reprogram their TCRs to target her cancer. Finally, the first hanging IV bag full of her virus-reprogrammed CAR-19 T cells was slowly infused back into Emily's veins.[14] By the third treatment, her side effects started.

Powerful cytokines triggered by the turbocharged immune onslaught whipsawed through Emily's system. At the time, physicians were not familiar with the extreme toxicity from the new T cell therapy,[15] but now they know it by several names; most scientifically "cytokine release syndrome" (or CRS), most notably "cytokine storm," most casually "shake and bake." As the names suggest, it is a whirlwind of exhausting and dangerous symptoms caused by the flood of signaling chemicals released during a T cell feeding frenzy, a monstrously amplified version of the immune side effects experienced during a debilitating battle with the flu. Emily's CRS was, in the language of her medical reports, "severe." Children have more powerful immune systems than adults; as the first pediatric CAR-T patient, Emily's CRS was more extreme than anyone could have anticipated. She was sweating and shaking and had trouble breathing, and her blood pressure dropped perilously low. As her temperature spiked to 105 degrees, Emily was rushed to the intensive care unit. She stayed there, a tube down her throat, another in her nose, breathing only by means of a ventilator. On the fifth day she was given steroids, which had been shown to sometimes lessen the severity of toxicity experienced in some anti-CTLA-4 patients. Emily's fever lessened temporarily, only to gain force like an offshore cyclone and come raging back. On day seven the little girl rising to the pump of a ventilator motor was as swollen as a hot water bottle, with multiple organ failure. It seemed the cure, rather than the disease, would kill her.

Desperate, her oncologist, Dr. Stephan Grupp,[16] pushed the lab to rush a wide battery of blood tests covering every immune-related molecule they could think of. When the bloods came back two hours later, two numbers stood out. Both her interferon gamma (INFγ) and interleukin-6 levels were remarkably high.

Grupp carried the readout into his 3 p.m. lab meeting to brainstorm the options. Nobody saw any. Emily's IL-6 was spiked a thousand times above normal levels, but the consensus was that was a red herring. IL-6 is a cytokine with a host of roles in normal immune function, both inflammatory and anti-inflammatory. It also happens to be partly responsible for the inflammation of rheumatoid arthritis.[17] And this was where Emily Whitehead got very lucky.

Dr. June was intimately familiar with the debilitating effects of rheumatoid arthritis in children. "One of my daughters has it," he explained. Her otherwise crippling disease was under control, but June had done his own research, and for several years had been following the progress of a promising new antibody that had been discovered to block the IL-6 receptors and shut down the cytokine call for inflammation and swelling. The antibody had been FDA approved for arthritis sufferers only a few months earlier under the name tocilizumab, and June had stocked up on it, just in case his daughter needed a backup therapy. "No one working in cancer would have had a reason to know about it," June said. "But it was just pure luck that I did." Now, June wondered, would this new arthritis drug help a kid with cancer?

It might, if the IL-6 was the cause of Emily Whitehead's cytokine storm. There were no experts to consult—they *were* the experts, and this was all unfamiliar territory. There was no time to lose; Emily's fever had hit 107 degrees, and her family had been told to be ready to consider a do-not-resuscitate order. Dr. Grupp wrote a script for tocilizumab,[18] ran down to where Emily was languishing in the ICU, and told the doctors what he planned to do. "They called him a cowboy," June remembers. The drug had never been tried against CRS patients, it had never even been proposed before. It was all brand-new.[19]

Grupp injected the tocilizumab into Emily's IV port. And gradually, the anti-IL-6 antibodies blocked their receptors and calmed Emily's cytokine storm. Over the next days Emily was weaned off the ventilator and the blood pressure medications, but she remained in a coma. A week later, she opened her eyes to the tune of "Happy Birthday" sung by the hospital staff. She was exactly seven years old and had made it to the other side.

~

CAR-Ts are robocop lymphocytes, what Dr. Sadelain calls "a living drug," and what Dr. June sometimes refers to as "serial killers" of cancer. A single CAR-T cell can take out as many as a hundred thousand cancer cells and produce freakishly rapid remissions that take even the most ardent immunotherapist by surprise. Only four weeks after the infusion Emily's biopsy result showed NED—a lab error, obviously, so Dr. June ordered a second biopsy. But there was no error. The procedure had been a success—as a drug for Emily, and as a proof of concept. That was good, but not the end. Emily wasn't the only childhood leukemia patient to receive the experimental treatment.

June had also treated another juvenile ALL patient at Children's Hospital, a ten-year-old girl. Her leukemia had responded to the CAR-T therapy, and she had gone into remission, only to relapse two months later. Biopsies showed that this girl's leukemia had mutated and escaped on B cells that did not carry the CD19 target protein. The cancer had changed uniforms. But they didn't have another CAR-T to give her. And so in September 2012, Emily Whitehead returned to school and become a national success story, a miracle featured on *Good Morning America*, and one of the symbols of hope that made the cancer moon shot possible. The other girl died of her cancer, a sad and humbling reminder of the work yet to be done.

~

Dr. Sadelain and the MSKCC group were the first to start CAR-19 T cell clinical trials, Dr. Rosenberg and his colleagues at the NCI had been the first to publish. Their successful CAR-T trial shrank tumors in a lymphoma patient.[20] That result was solid, but not nearly as dramatic as the complete success in a childhood cancer patient. That made the headlines and energized the entire field, propelling CAR-T funding and development into overdrive. Each of the teams—once collaborators, now competitors—quickly coupled up with a pharmaceutical partner to turn the technology into medicine. The National Cancer Institute went with Kite Pharma[21] (which gained approval for its CAR-T, called Yescarta,[22] for large B-cell lymphoma); Memorial Sloan Kettering Cancer Center, together with the Fred Hutchinson Cancer Research Center and the Seattle Children's Research Group, partnered with Juno Therapeutics. The drug giant Novartis licensed the CAR-T technology from the University of Pennsylvania, and received FDA approval for the therapy used on Emily Whitehead, which it sells under the brand name Kymriah. That approval only came in 2017, but CD-19 CAR-T therapy has already helped thousands of people, including more than one hundred children with cancer, making it one of the best examples of how rapidly immunotherapies are changing our relationship with the disease.

~

Kymriah is both a medicine and a product. It is dispensed from a handsome translucent package with a blood-orange luminosity. Each is customized for the patient, engineered from the patient's own T-cells. At the moment each of these bespoke one-time infusions cost $475,000. When hospital charges are added, the total cost approaches a million dollars per patient. The next best treatment for acute B-lymphoma is a bone marrow transplant costing over $100,000 more. (This "economic toxicity" is currently another serious side effect of cutting-edge treatments like cancer immunotherapy. Whether these prices are correct, fair, or tenable is a worthy question, beyond the scope of this book.)

For a CAR-T patient, the process of receiving the treatment goes something like this: An eligible patient—often a kid with otherwise untreatable lymphoma—travels to a medical center, certified by Novartis in the procedure. (As of February 2018 there are twenty-three treatment center sites across the United States.) There, blood is drawn from the patient and centrifuged for at least fifteen minutes at 2200–2500 rpm to separate the T cells from the plasma, platelets, and the rest. The T cells are then cryogenically frozen, packed in a special cryovac container, and shipped off to the 180,000-square-foot Novartis mothership facility in Morris Plains, New Jersey, where they are thawed and reengineered to recognize a protein specific to the patient's cancer. This proceeds in steps. First the T cells are activated. Then they are transduced with a virus containing new genetic instructions. Then they are grown and multiplied until they number in the hundreds of millions. The clone army of supersoldier T cells is then re-cryopreserved, shipped back to the certified medical center, and rethawed to be dripped back into the patient.

Cryopreservation allows patients from all over the world to use the treatment. The turnaround time, from walk-in center to finished customized T cell treatment, is twenty-two days. Preliminary data suggests that therapies using these bespoke T cells deliver durable response rates for formerly hopeless cases.

Emily is part of that happy statistic. As of August 2018 she remains in remission.

There is great danger, of course, in any engineered fiddling with the hair triggers, feedback loops, and checks and balances of an immune system evolved over millennia, and great trepidation in using experimental therapies on any patient, especially a child. At the same time, the worst possible side effect of these treatments is death; leukemia ends the same way without the treatment. Those first experimental treatments, and the new approach to treating cytokine storm, quickly demonstrated that for these patients, the rewards far outweighed the risks. For such patients, CAR-T has changed the numbers seemingly overnight. In the ALL group that had formerly a zero percent survival rate, estimated survival rates now stand at 83 percent or higher. Large B cell lymphoma was the next target for CAR-T, and now more are in the works, with several already in clinical trials. Those new targets include leukemia, chronic lymphocytic leukemia, multiple myeloma, recurrent glioblastoma, advanced ovarian cancer, and mesothelioma. Solid tumors still remain a challenge, but this technology is as new as it is powerful, and it is evolving extremely quickly, with several spin-off companies chasing versions of CAR-T that use donated T cells for off-the-shelf solutions. Others are programming "kill switches" into the CAR-T cells, so if this Frankenstein goes crazy we just pull the plug. CAR-T is so powerful and so new—the first approval was only in 2017—it's impossible to imagine what another year will bring. But whatever it is, we can now hope that patients like Emily Whitehead will be around to see it.

Chapter Eight

After the Gold Rush

The new cancer drugs with four-syllable names are now products sold during Super Bowl ads, and, remarkably, the new Jimmy Carter drug isn't new anymore. But the surprise, excitement, and hope surrounding that first breakthrough in cancer immunotherapy triggered a flood of new interest and funding to the field, and a multiplier effect for the pace of scientific progress. The result is what biologist E. O. Wilson has referred to as a "consilience"—an intellectual synergy that comes when specialists from very different disciplines are able to examine a common topic and find a common language to share their ideas. It's no longer an argument between cell biologists and immunologists and virologists and oncologists; it's a conversation. For the first time we all glimpse the whole of the cancer-immunity cycle. The blind men and women examining the elephant have suddenly gained sight and can get to work.

The result is billions of dollars and scores of talented specialists now devoted to cancer immunotherapy. The funding torchbearers of the field such the Cancer Research Institute, started more than seventy years ago by William Coley's daughter, have been joined by new organizational infrastructures to support that work, among them the Biden "moon shot" Cancer Initiative, to rethink medicine as a whole, and cancer most specifically; the Parker Institute for Cancer

Immunotherapy to fund and coordinate researchers and clinical trials as never before; public appeal drives such as Stand Up to Cancer; (SU2C), which directs hundreds of millions of donated dollars directly into research and clinical trials; and a gold rush for commercial pharmaceutical companies and startups and the dozens of biotech venture capitalists that fund them. Several researchers have quipped that there are now two types of drug companies: those that are deep into cancer immunotherapy, and those that want to be.

For everyone—the organizations, the individuals, and most especially the patients—the goal is to change what it means to have cancer and to render the disease a chronic condition, serious but manageable, like diabetes or high blood pressure. Or, just maybe, to cure it.

Cure isn't a word thrown around lightly by oncologists, but now the top scientists in the cancer field are willing to bandy the word about aloud, publicly and often. In fact, they remind us, we already have cured cancer in a subset of patients. The work now is to enlarge that cohort. The classes of cancer immunotherapies described below are among the ones that might help accomplish that goal.

∼

Checkpoint inhibitors may be the purest articulation of cancer immunotherapy, because they simply unleash the immune system. The first was the anti-CTLA-4 antibody ipilimumab, which won FDA approval for metastatic melanoma in 2011.[1]

That drug was an immediate game changer, reducing deaths from late-stage melanoma by 28 to 38 percent. The first phase 1 clinical trials started in 2001, long ago enough to qualify 20 to 25 percent of those patients with "long-term survival" benefit. That's still less than half of the patients, but a great deal better than the low single-digit survivor percentages only the year before.

Dr. William Coley (center) and colleagues. *(Cancer Research Institute)*

"Mr. Zola." *(Cancer Research Institute)*

Investigators and Medarex and BMS employees, 2010 American Society of Cancer Oncology presentation of Phase 3 anti-PD-1 antibody data (standing from left): Jedd Wolchok, Jeff Sosman, Jeff Weber, Axel Hoos, Geoffrey Nichol, Israel Lowy, Mike Yellin, and Alan Korman; (kneeling from left) Stephen Hodi and Nils Lonberg. *(Axel Hoos)*

(Left to right) Dr. Carl June, Kari Whitehead, and Tom Whitehead, with their daughter, Emily, the first pediatric cancer patient to receive CAR-T therapy. *(Cancer Research Institute)*

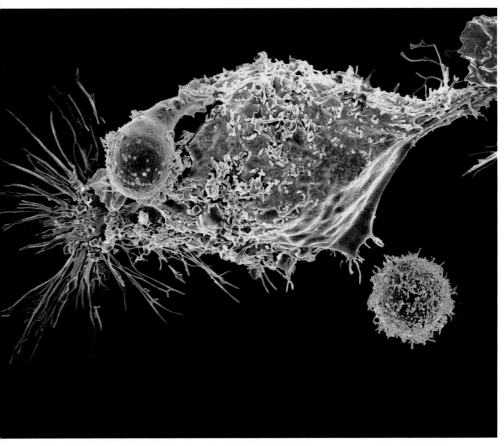

Engineered T cells (CAR-T) attacking a cancer cell. *(Prasad Adusumilli, MSKCC)*

Brad MacMillin celebrating his 2-year NED anniversary with Dr. Dan Chen. *(Emily MacMillin)*

Emily MacMillin: "8/30/2009. Erin (our third) born, with Clare and Camille, during a time when we thought we had beaten the odds (again)." *(Emily MacMillin)*

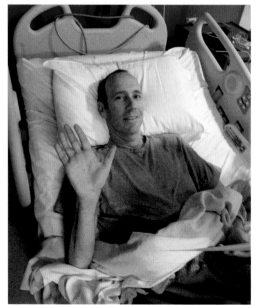

Brad MacMillin, July 2013, at MD Anderson Cancer Center. *(Emily MacMillin)*

Kim White: "Dan and I from the benefit concert he did for me. July 17, 2014." *(Kim White)*

Jeff Schwartz and Kim White. *(Kim White)*

Kim and her husband, Treagen, with their daughter, Hensleigh. *(Kim White)*

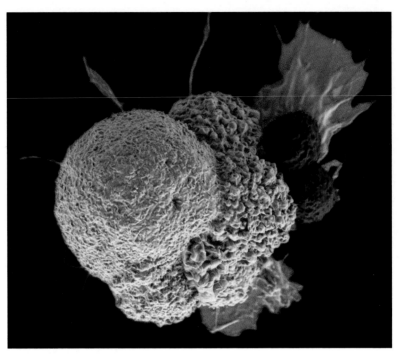

Colored scanning electron micrograph of a cancer cell (white) attacked by cytotoxic "killer" T cells (red). *(National Cancer Institute\Duncan Comprehensive Cancer Center at Baylor College of Medicine, Rita Elena Serda)*

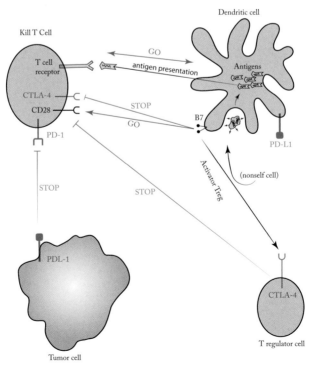

T cell activation and some checkpoints (CTLA-4, PD-1/ PD-L1) known to stop it

Anti-CTLA-4 drugs have some serious toxic side effects, but they set the stage for other immunotherapies, including more selective checkpoint inhibitors, such as the anti-PD-1 / PD-L1 drugs.

There are now at least half a dozen approved anti-PD-1 / PD-L1 drugs.[2] Each blocks one or the other side of the handshake. Whether it matters which side of the handshake you block, only further trials will tell. The anti-PD-1 / PD-L1 drugs seem to work best if a patient's tumor is expressing PD-L1. For that subset of patients, the drug has worked well, providing durable and sometimes complete responses.[3]

Both types of checkpoint inhibitors prevent cancer from turning down or off the immune response, but there are important differences between these drugs, having to do with when cancer uses the checkpoints they inhibit. CTLA-4 is a more general checkpoint; it happens earlier, preventing T cell activation, and when you block it, the response may be more general, too.[4] Cancer uses the PD-1 / PD-L1 checkpoint later, after the T cell is activated. Blocking these checkpoints has a more specific response, like removing handcuffs only from specialized soldiers already on the battlefield and face-to-face with the enemy. As you might expect, the anti-PD checkpoint inhibitors are better tolerated and have far fewer toxic side effects than anti-CTLA-4, which is now understood to both up-regulate T cell activity and down-regulate the specialized regulatory T cells, or T regs, that keep the immune system from overreacting.

Both of these drugs—but especially the anti-PDs—are proving to be even more effective in *combination* with other therapies. As the data rolls in, it seems that most cancer therapies work better when coupled with a PD-1 / PD-L1 checkpoint inhibitor. That includes chemotherapy, which kicks off the immune battle with some dead tumors for the unleashed T cells to recognize and activate against.

For example, over the course of one week in July 2018, data from phase III trials showed that a combination of an anti-PD-L1 drug and a chemotherapy agent showed meaningful improvements against both small-cell lung cancer and triple-negative breast cancer—the first advances against either disease in decades.

Cancer immunotherapies that previously failed are now being reevaluated to see if they work better with the brakes off (i.e., in combination with a checkpoint inhibitor). Most of those combinations are with anti-PD-1 / PD-L1 drugs. And future cancer therapies are now being planned with checkpoint inhibition in mind. The net result is that most drug companies with cancer therapies in their portfolio want a PD drug to pair it with. There are now reportedly *164 PD-1 / PD-L1 drugs* in the pipeline between preclinical testing and consumer marketing, and industry insiders suspect there may be many more being developed in China. This redundancy isn't the best use of intellectual or physical resources; one hopes it will result in more competition and lower prices.

(A question not addressed in this book is how anyone's going to afford this bright shiny future. Pricing for Yervoy—the trade name for the anti-CLTA-4 drug ipilimumab—is typical, costing more than $120,000 for a four-course treatment. Merck's anti-PD-1 drug Keytruda, for advanced melanoma, costs $150,000 for a yearlong treatment. Behind the good news is a pressing need for better answers about how we pay for the inevitability of illness and decline. Cancer is an equal opportunity disease; if progress against it isn't for everyone, even a breakthrough becomes a step back for our humanity.)

~

When I ask immunology researchers what's next, the answer is always "more": more tools, more targets, more therapeutic agents.

More drugs, more FDA approvals and fast tracks, more *biomarkers* to better describe cancer with molecular specificity (as opposed to classifying it by the organ the mutation started in, be it liver, lung, or breast), combined with more "immune profiling" of the specifics of a patient's immune system (to determine who will receive the greatest benefit from exactly what type of immunotherapy).

Such *personalized cancer immunotherapy*, which pairs an individual's unique immune profile and unique tumor genotype with the right immunotherapy combination, is presumed to be the future of cancer treatment.[5]

It's not reasonable to guess at what will work next, but as of this writing the greatest demonstrated promise—the closest to a sure thing in terms of what's been demonstrated in the clinic—seems to come from the expanding CAR-T therapies and CD3 bispecifics. Watch this space. It's moving fast.

As of June 2018, there were reported to be some 940 new immuno-oncological drugs being tested for breakthrough designation and FDA approval. Another 1,064 new immunotherapy drugs are in the labs in preclinical phase.

That's 2,004 new cancer drugs in just a few short years. This speed of change is highly unusual in medicine, and totally unprecedented in cancer. And by the time you read this, those numbers and the science behind them will have advanced again.

∽

It's worth noting that a study published in the *Proceedings of the National Academy of Sciences of the United States of America* found that every single one of the drug approvals since 2010—210 of them—can be traced back to the $100 billion NIH budget for drug development. The breakthrough is built on tax dollars, and it's yours.

Chapter Nine

It's Time

Jeff Schwartz was feeling better during the early summer of 2014, which was good for a number of reasons, including that one of his bigger acts was touring. Imagine Dragons is a four-piece indie rock band fronted by a young guy named Dan Reynolds—they were all young, Jeff says, young and friendly and impossibly nice and all deeply religious. They had all met in their native Salt Lake City, Utah. They were talented and, suddenly, they were explosively popular—the biggest musical act of 2017, according to *Billboard* magazine—and yet they hadn't let the rock star experience go to their heads. That was unusual, but it was beautiful, too.

For members of the Church of Jesus Christ of Latter-day Saints, service is an obligation, not a luxury, and charity was one of the core principles of the band. One of the first charities they established was dedicated to a young fan named Tyler Robinson. Tyler had been diagnosed with a rare form of soft tissue cancer called rhabdomyosarcoma, stage 4. Tyler's older brother had written a letter to the band, telling them Tyler was sick and would be in the audience for their show in Provo, Utah. Tyler had found inspiration in the band's music, he'd said, especially their song "It's Time." The lyrics "the path to heaven runs through miles of clouded hell" resonated for him and kept him running those miles. Could they send him a shout-out of encouragement, say something to cheer him from the stage? Tyler's brother said he'd be hard to miss. He'd be the bald sixteen-year-old, frail enough to be carried on his big brother's shoulders.

There's some raw video of that night, and if you aren't one of the over 600 million people who have watched it on YouTube alone, you should. It's inspiring and beautiful and hard all at once.[1] The video is shot on a cell phone. The room is intimate; we're in the crowd. They are excited and young, local kids shoulder to shoulder at a singalong with their home-state heroes. The band is close, bathed in red spotlights and milling between songs as lead singer Dan Reynolds takes the microphone stand.

"I've got something serious to say—it'll be the only serious thing of the night, I promise," he says, "so—if you could pay attention, and a little bit of quiet please, this is something really important to me." That's no easy request after the previous hour spent whipping the kids into an inspired frenzy, but this is their hero speaking, rubbing his head and pacing, and they listen as he calls out the name of someone in the crowd: "Tyler Robinson." Dan had read Tyler's story just that night. "I want to say that it very much inspired me. It meant a great deal…" The crowd gets a notch quieter when he says the word: *cancer.*

"And, um, he asked us to play a song for him tonight, and this song is for Tyler—from the bottom of our hearts, Tyler." Dan touches his heart under the sweaty white T-shirt and lifts a fist in the air in solidarity. It's a signal to the crowd. They whoop and yell in affirmation as the band kicks into the familiar strains of their megaplatinum hit. There's a sudden urgent scrum around the pale, bald white kid in a blue T-shirt, hugs and *I love you man*s and a blurry push to get him toward the front of the stage, where Dan is singing the words, words everyone knows. Now Tyler is close and thrilled and singing them back, pointing to punctuate every line of failure and affirmation:

And now it's time to build from the bottom of the pit
Right to the top
Don't hold back

Dan leans in. He grabs Tyler's hand as his brother gets underneath and lifts him onto his shoulders, still singing, tall and hit by the spotlights, and the crowd screams in understanding—*that's the kid!*—just as the song builds to the sing-along refrain. Tyler shouts the words, believing them. "It's Time" is a swelling, inspiring song that builds and breaks and becomes an anthem. Tyler was about to finish chemo, about to maybe be just a kid again rather than the cancer kid, and though he's not there yet, he's *here*, owning the lyrics because they'd spoken to him personally, as a friend, the way music does, especially when you're sixteen years old, especially in heartbreaking times. These words were about his own life, and when he sang "the path to heaven runs through miles of clouded hell" he was being literal and specific. The crowd seems to understand that, as Tyler and Dan lock together in the spotlight, preaching "It's time to begin" head-to-head in a sweet soul kiss blessed by a thousand parishioners. And you can tell, even on the shaking video from a camera phone, that the band is more surprised than anyone, that the song is no longer theirs, that the lyrics are no longer Dan's, and that it's this grinning, confident, beautiful, bald, dying kid they'd met only a minute before who owns it.

This was one of those moments Jeff Schwartz talks about, the thing only live music does, the melting snowflake moment, electricity that's there and then not. The song ends, but the crowd won't let the moment go. They spark, then explode once more, chanting *Ty-ler! Ty-ler!*, an echo of a moment too precious to die.

Tyler and the band kept in touch after the show, and when Tyler was declared cancer-free at the end of 2011, it seemed like the end of what they'd begun in that auditorium, a fulfillment of something prophetic and empowering. Because, finally, it *was* time, and Tyler could get back to being a teenager.

But not all beautiful stories have happy endings. In March 2013, Tyler fell into a coma with a brain full of cancer; by spring he was dead. It was a shock, an especially mean turn after what felt like victory, and maybe even harder for the millions of young Imagine Dragons fans who had seen music and youthful will defeat this old-people disease and hadn't ever had to experience the cruel and too common opposite. It's always shocking when young people die, and it's especially shocking to other young people. This was an extra mile of clouded hell too many.

The band crafts catchy hooks and rousing millennial anthems. They couldn't cure cancer, but they could use their success to help those dealing with it, and create a foundation in Tyler's name to help soften the financial blow for the families,[2] a foundation their cancer-surviving accountant Jeff could help sort out. The entire band supported it, playing benefits together or separately, depending on their schedule. In July 2014, the band was on a European tour, so only Dan Reynolds could fly back to Utah for an intimate one-night acoustic benefit raising money for another young cancer patient. This time, it was someone he'd met as a teenager eight years earlier at a Latter-day Saints fireside youth mixer back in Salt Lake.[3]

~

Kim White[4] was a pretty young member of the extended Imagine Dragons family, a mom with a young child and another on the way. She'd been having inexplicable blood pressure problems in the second pregnancy and when medication didn't help, her obstetrician ordered an ultrasound and discovered an eleven-centimeter tumor wrapped like a boxing glove around the adrenal gland above her right kidney. A battery of tests convinced her doctors that the tumor was benign and that both it and her pregnancy could be managed safely.

But four weeks later Kim developed HELLP syndrome, a poorly understood disease associated with pregnancy that includes symptoms like preeclampsia. Kim was hurriedly prepped for emergency surgery to remove the tumor and deliver the baby immediately. There was no chance her eighteen-week-old son would survive outside the womb. If they didn't do the surgery, both would die.

Like the band, Kim and her husband, Treagen, were active members of the Church of Jesus Christ of Latter-day Saints. They now leaned hard on that faith. As Kim's father and husband lay their hands on her head to deliver a blessing for comfort, healing, and counsel, Kim felt Jesus Christ's presence. "He was wrapping his arms around both of us," she said. "Promising us that everything was going to be OK."

But after the surgery, Kim's diagnosis was anything but OK. The tumor *was* cancer, an unusual and aggressive form called adrenal cortical carcinoma. Kim's cancer was stage 4. Her doctor told her that she'd be lucky to survive five years.

Kim's oncologist started her on chemotherapy. She cut off her long, corn-silk-blond hair, once a key aspect of her stand-out beauty, and tried to care for her eighteen-month-old daughter without thinking that she'd likely soon be without a mother. She felt sadness and fear, but mostly what gripped her was white-hot rage. "I spent many nights in the bathtub screaming at the top of my lungs about how angry I was, how unfair life was, how could you let this happen to me," she wrote. "Trust me, if you've thought it, I've said it."

An LDS counselor reminded her that whatever she was going through, Jesus Christ had gone through it already. Nobody else might understand how unfair it was, but He did. And so Kim doubled down on her relationship with her savior, and her faith in both a higher power and her oncologist. And she made a conscious decision to start counting her blessings and look for something positive in each day she had left. "All that anger wasn't helping me any," Kim said. "People don't realize how mental a physical battle is."

As the bills started mounting up, a friend set up a GoFundMe account to raise $10,000[5] to cover another surgery, the first of fifteen. The larger Imagine Dragons family started chipping in as well. When Dan heard about it, he decided to do the July 14 benefit concert.[6] Jeff Schwartz would help figure out the money part. The benefit gave Kim hope and inspiration, the way the concert had given Tyler Robinson hope. And it raised some $40,000 toward her medical bills.

But when Kim went in for her next scan, her oncologist came back with devastating news. The tumors were spreading, metastasizing throughout her body. They were especially prevalent now in her lungs. The oncologist counted with an eraser: nearly fifty tumors. "My oncologist was like, look, [the chemotherapy] isn't working, and it's making you so sick," Kim said. "I've called other doctors and I'm at a loss for what else to do." Kim told her husband, and her father, and a small group of her friends. She told Dan. And then she told Jeff Schwartz.

Jeff wanted Kim to have the same opportunity he had, to walk through the same door that he had. A door in LA that happened to open into one of the few medical centers in the country then conducting human trials on the new cancer immunotherapy drugs. She could go there. It had worked for him. At the very least, it was worth a try.

"Her own doctor in Salt Lake City, he kinda pooh-poohed it at first," Jeff says. He hadn't seen the rare form of cancer Kim had, but he'd seen plenty of patients jump at the promise of new miracle cures over the years. A young patient talking about some experimental miracle medicine she'd heard about from some guy, a friend's accountant—that was cruel folly this poor girl didn't have time for. The fact that this guy was proposing immunotherapy made it even worse. It didn't work—he'd been to the cancer conferences, he knew.

Nevertheless, Kim prayed and thought and decided, once again, that she really didn't have a choice. The chemotherapy wasn't working. Soon, she'd be too weak to even make this choice. And so she followed Jeff's advice. She went to LA and walked through the door.[7]

Kim met with Jeff's doctor at the Angeles Clinic and was able to enroll in a study for a checkpoint inhibitor called pembrolizumab[8]— "the Jimmy Carter drug."[9] It wasn't the same drug Jeff had gotten, but it was closely related, a PD-1 / PD-L1 checkpoint inhibitor, just the other (PD-1) side of the handshake. Jeff's drug blocked the receptor on the tumor. Kim's drug did the same thing on the T cell side.

The Angeles Clinic had been one of the original fifteen sites where the anti-PD-1 drug had undergone clinical trials[10] and was one of the first places where the drug, which the FDA had deemed a "breakthrough" therapy, was made available to patients. It didn't work for all cancers or all patients, but Kim had nothing to lose, and nothing left to try. "So we just had to go for it."

Jeff next saw her after she had received her first doses of the drug. She was still frail and thin, but she seemed to have gained a little weight back. "I'm looking at her," Jeff says. "Her eyes were bright. She looked scared. But she also looked better.

"I tell her, 'Hey, the tumors are melting away, I know it,'" he says. "How do I know? I don't know." He showed this sort of confidence with his clients sometimes, they believed him, and that confidence made a difference. But show biz isn't science. Jeff's prognosis was pure hope and boosterism.

"I was trying to keep her spirits up," Jeff says. "Keep her going back." But if it was really working—of course, he had no idea.

When Kim got her CT scan a few weeks later, she called him, in tears.

"I heard the crying—I was worried," Jeff says. "Then she told me, 'I had forty-two tumors in my lung—now I have two.' 'I had...'"

Jeff listened as she continued the litany. "So, I'm Superman, I saved her life," he says, and as he laughs now he's laughing both at himself and at the truth of that truth, too. What saved her life was hope and information.

"Kim wanted to stop being a sick person, leave this cancer chapter behind her, and get back to her life. So I got sorta angry with her," Jeff told me. "I said, 'No, you're not done. Now what you gotta do, you gotta pay it forward. You got lucky. Now you've got to make sure other people have the chance to get lucky like that, too.'"

Paying it forward, sharing information, and telling stories— these are sentiments common among survivors of cancer, and of those who lost loved ones. That's why Emily shared Brad's story with me. It's a form of gratitude—for what Dan Chen did for her husband, what all her doctors did, or tried to do. And a hope that others can learn from her story and perhaps have better outcomes. That was what Jeff was impressing upon Kim. She had been given an opportunity, not just to live but to share her story, as he'd shared his. And she could share it more widely.

"I told you, she's a Mormon," Jeff says. "The Mormons, they can post a fart on Facebook, they get, like, a hundred thousand thumbs-up, it's an incredible network they've got. I tell her, 'You've got to use that, and make sure other people know what happened to you. Tell them, you had this rare cancer. You used this drug, you had this experience.'"

And so, instead of just resuming her life as one of the lucky ones, Kim started a foundation, spreading awareness about the specifics of her cancer and the way it responds to the new generation of immunotherapy drugs. And that's why she is willing to share her story here. Her health issues weren't over. She's been through a lot since then. "This isn't just about that drug, or about immunotherapy," Kim says. "There's a lot more to it. I've gotten so much support from my family and friends, from strangers going through the same thing. And, wanting to be there for my daughter, that kept me moving, too. The attitude you have, the mental part, staying positive, it's maybe the most important part." The drug didn't cure her, not completely. "But that drug saved my life," Kim says. "If it wasn't for that, I wouldn't be able to even try and fight all this other stuff."[11] This is her service, her act of faith. Now she's helping save lives, too.

~

If you look back at any series of events, try to break it down to all the unlikely little details that had to line up for one outcome to happen, everything looks like a miracle. The mutated cell, the missed flight on 9/11, the one seat left in the movie theater next to a stranger who becomes your husband. It's all statistically improbable. It's also all inevitable.

But the truth is, if Kim hadn't gone to a church bonfire youth thing as a teenager and met a guy who became a rock star; if the rock star hadn't dropped out of college and written a song that inspired a cancer kid named Tyler Robinson, whose death inspired the band; if Jeff Schwartz had gotten Yankees tickets instead of Mets tickets, if he hadn't given them away, gotten the job, and repped Joan Jett; if Dan Chen hadn't decided to put Jeff on the study, and if it hadn't been Christmas, or if the phone call hadn't gone through and they'd started Jeff on the other study; if a young surgeon named William Coley hadn't felt so badly about another pretty young cancer patient that he couldn't save, if he hadn't gone searching the hospital records for stories of miracles and spontaneous cures—really, there's no end of ifs. In the end, Kim got lucky. Her own cancer doctor didn't know about the drug trials Jeff had been in. We don't need the miraculous here. We have a rock-and-roll accountant advising a young Mormon mom on a breakthrough cancer treatment. Whatever that is, a lucky break, the unseen hand, a higher power, or just history, it doesn't matter. It's her story now, and yours. Pass it on.

Acknowledgments

Most of the people responsible for cancer immunotherapy are missing from this book.

Stories love a hero, but science doesn't really work that way. By necessity, this science story shines a penlight on a handful of individuals. The bulk of the major characters sit more quietly, in the endnotes and appendix; others are missing entirely. They are not a supporting cast but the cast itself, and yet if each were introduced this book would be unreadable. Even less heralded are the countless patients who volunteered their lives. If we're looking for heroes, we start with the authors listed in the academic papers, and the patients who are not. Another important acknowledgment is that every breakthrough is built on the ruins of shattered certainties. Those in this book will also surely change as research continues. It's changing quickly.

This book was made possible in part by the vision and generosity of the Alfred P. Sloan Foundation. Special thanks to Doron Weber, Program Director for Public Understanding of Science, Technology & Economics, who knows writing as only a great author can, and loss as only a father can. He's a true mensch and champion of the arts.

Many of the people in this book are literally curing cancer yet somehow found time to let me bother them repeatedly. This book is possible only because of their generous patience and tutelage. Visionary publisher Sean Desmond bravely took on this project and the years required. Susan Golomb somehow helped find a home for a breakthrough few had heard of; her father, Dr. Frederick M. Golomb, had experienced the confounding failure of immunotherapies in the 1950s and 60s and so was my first skeptic. Thanks for the above and beyond from Matt Tontonez; Brian Brewer; Mary Riner; Rachel Kambury; Adam Piore; Dragon Yung Mei Tang; Starvos Polentas; Michael "The Wheel" Lafortune; Pete Mulvihill; Cary Goldstein; Dr. Kauffman, J.D.; Mac Reynolds; Arun Divakaruni; Matt, Jana, and Master Thomas in Farmington; Margaret Van Cleve, io360, and Julia Gunther at AACR; Nick and Caroline, who kept spirits bright; the patient and kind Bob Castillo; the ultimate immunoeditor, the legendary Ann Patty; and Dr. Charles W. Graeber, who turns good students into great physicians and is as fine a doctor, friend, and father as one could hope for. Diann Graeber, you already got a dedication so I know this makes you laugh. Special appreciation for the patients and their families, named or otherwise, whose deeply personal stories are no small thing to trust a stranger with. I am grateful to the Virginia Center for the Creative Arts and the inspiring visits of Elizabeth and Sabine Wood during my seasons of seclusion, as well as The Writers Room NYC, Java Studios Greenpoint, and the Grey Lady in winter. The often patient Gabrielle Allen truly makes all things possible. Thanks and celebration to her family for the long and Mainey life of Dr. Tom T. Allen, beloved pugilist, sailor, and osteopath. In remembrance and gratitude for the extraordinary life of Camilla Sewall Wood, and to the Baldwin family for the inconsolable loss of Malcolm. In memory of John P. Kauffman, 1971–2018, a really great guy gone too soon.

Cancer Research Institute William P. Coley Award Recipients

2017: Rafi Ahmed, PhD; Thomas F. Gajewski, MD, PhD. **2016:** Ton N. Schumacher, PhD; Dan R. Littman, MD, PhD. **2015:** Glenn Dranoff, MD; Alexander Y. Rudensky, PhD. **2014:** Tasuku Honjo, MD, PhD; Lieping Chen, MD, PhD; Arlene Sharpe, MD, PhD; Gordon Freeman, PhD. **2013:** Michael B. Karin, PhD. **2012:** Richard A. Flavell, PhD, FRS; Laurie H. Glimcher, MD; Kenneth M. Murphy, MD, PhD; Carl H. June, MD; Michel Sadelain, MD, PhD. **2011:** Philip D. Greenberg, MD; Steven A. Rosenberg, MD, PhD. **2010:** Haruo Ohtani, MD; Wolf Hervé Fridman, MD, PhD; Jérôme Galon, PhD. **2009:** Cornelis J.M. Melief, MD, PhD; Frederick W. Alt, PhD; Klaus Rajewsky, MD. **2008:** Michael J. Bevan, PhD, FRS. **2007:** Jeffrey V. Ravetch, MD, PhD. **2006:** Shizuo Akida, MD, PhD; Bruce A. Beutler, MD; Ian H. Frazer, MD; Harald zur Hausen, MD. **2005:** James P. Allison, PhD. **2004:** Shimon Sakaguchi, MD, PhD; Ethan M. Shevach, MD. **2003:** Jules A. Hoffmann, PhD; Bruno Lemaitre, PhD; Charles A. Janeway Jr., MD; Ruslan Medzhitov, PhD. **2002:** Lewis L. Lanier, PhD; David H. Raulet, PhD; Mark John Smyth, PhD. **2001:** Robert D. Schreiber, PhD. **2000:** Mark M. Davis, PhD; Michael G. M. Pfreundschuh, MD. **1999:** Richard A. Lerner, MD; Greg Winter, PhD; James E. Darnell Jr., MD; Ian M. Kerr, PhD, FRS; George R. Stark, PhD. **1998:** Klas Kärre, MD, PhD; Lorenzo Moretta, MD; Ralph M. Steinman, MD. **1997:** Robert L. Coffman, PhD; Tim R. Mosmann, PhD; Stuart F. Schlossman, MD. **1996:** Giorgio Trinchieri, MD. **1995:** Timothy A. Springer, PhD; Malcolm A. S. Moore, PhD; Ferdy J. Lejeune, MD, PhD. **1993:** Pamela Bjorkman, PhD; Jack Strominger, MD; Don Wiley, PhD; John Kappler, PhD; Phillippa Marrack, PhD; Alvaro Morales,

MD, FRCSC, FACS. **1989:** Howard Grey, MD; Alain Townsend, PhD; Emil R. Unanue, MD, PhD. **1987:** Thierry Boon, PhD; Rolf M. Zinkernagel, MD, PhD. **1983:** Richard K. Gershon, MD. **1979:** Yuang-yun Chu, MD; Zongtang Sun, MD; Zhao-you Tang, MD. **1978:** Howard B. Andervont, PhD; Jacob Furth, MD; Margaret C. Green, PhD; Earl L. Green, PhD; Walter E. Heston, PhD; Clarence C. Little, PhD; George D. Snell, PhD; Leonell C. Strong, PhD. **1975:** Garry I. Abelev, MD, PhD; Edward A. Boyse, MD; Edgar J. Foley; Robert A. Good, MD, PhD; Peter A. Gorer, FRS; Ludwik Gross, MD; Gertruda Henle, MD; Werner Henle, MD; Robert J. Huebner, MD; Edmund Klein, MD; Eva Klein, MD; Georg Klein, MD, PhD; Donald L. Morton, MD; Lloyd J. Old, MD; Richmond T. Prehn, MD; Hans O. Sjogren, PhD.

CRI Lloyd J. Old Award Recipients

2018: Antoni Ribas MD, PhD. **2017:** Olivera J. Finn, PhD. **2016:** Ronald Levy, MD. **2015:** Carl H. June, MD. **2014:** Robert D. Schreiber, PhD. **2013:** James P. Allison, PhD.

Appendix A

Types of Immunotherapies Now and Upcoming

The menu can be confusing, and it's changing.[1] It has changed dramatically in the course of the years during which this book was researched and written, and it will continue to change. But what can be helpful to keep in mind is that what most (not all) immunotherapies have in common is the T cell.

IL-2 grows and energizes them, adoptive T cell therapy grows and farms them, checkpoint inhibitors unleash them, vaccines inform and activate them, and CAR-T *is* them, the robocop version. Immune response is complex. There are many players involved, many are surely undiscovered, and only a few are understood. But in terms of cancer treatment, the goal is simple: get cancer-killing cells to do their job as quickly and selectively as possible.

Anything that accomplishes that is an immunotherapy.

That includes a class of immunotherapies that simply works as a universal adaptor on the molecular scale, chaining T cells (or natural killer cells) to the cancer cells with protein handcuffs. Called *bispecific antibodies*, or BsAbs, these bioengineering marvels are like an aggressive matchmaker at a high school dance. There's currently hope that such an approach will be more effective after a checkpoint inhibitor like anti-PD-1 / PD-L1 turns on the lights, revealing to the T cell that its dance partner is cancer.[2] Currently the CD3 bispecifics are particularly promising. These bind to the cytotoxic T cell–stimulating CD3 site on the T cell and various antigen targets on tumor cells.[3] Two such drugs have been approved by the FDA (blinatumomab, Amgen) and for use in Europe (catumaxomab, Trion Pharma). It's reported that more than sixty such drugs are in preclinical phase and thirty are in clinical trials, the majority of which have cancer as the target.

We are living in the checkpoint inhibitor phase of cancer immunology, or perhaps the second half of that phase (CTLA-4 was the pioneer, PD-1 / PD-L1 is the present), and already researchers speculate that we've plucked the low hanging fruit. This is the era of the combinations.

Combinations

In addition to combining existing checkpoint inhibitors with one another[4] (Ipi + a PD-1 / PD-L1), some combinations include checkpoint inhibitors plus: *chemotherapy*; *radiation therapy*; *T cell agonist cytokines* such as IL-2; new customized *vaccines*; and, in a technological twist reminiscent of Coley's Toxins, *inoculated bacterium* such as listeria or small molecules. This is a far from complete list.

There are more potential checkpoints being discovered, as well as numerous new therapeutic approaches to induce tumors that are not very immunogenic (visible to the immune system), to express unique antigens, or in some other way make those cancers viable immune system targets.

Anything that makes cancer more visible as an immune target is a potential partner for drugs that unleash immune cells to attack those targets.

The list of combinations being tried as of September 2018 is reported to run into the thousands.

Cellular Therapies

A "cellular therapy" is any cancer treatment that uses a whole living cell as the "drug" (rather than just a folded protein or other molecule as the therapeutic agent). That includes *adoptive T cell transfer*, a method that essentially farms T cells, grows out the ones effective against cancer, and transfers them back to the patient. Notable advances to this method were led early on by pioneering work from Fred Hutchinson Cancer Research Center's Phil Greenberg as well as work by Dr. Steven Rosenberg's colleagues at the National Cancer Center, which was one of the first centers to push this technique into the clinic and which has continued to grind out progress in this approach for decades. (In June 2018 Rosenberg's team published results of a successful adoptive T cell transfer therapy that saved the life of a forty-nine-year-old Florida woman with stage 4 breast cancer and large tumors throughout her body. As of June 2018 she had no evidence of the disease, after receiving an infusion of some 90 billion of her own T cells.[5])

CAR-T is (at present) the best-known cellular therapy and one of the most exciting to watch. It works. It's shown to be hugely effective against those cancers the CAR can be reengineered to target. Right now that's a limited subset of cancers, mostly blood-borne cancers. Various new approaches currently under way attempt to make that list of applicable cancers bigger, expanding the settings in which patients may safely receive this treatment while shrinking the price tag for what is now an entirely bespoke drug.[6] Advances in gene editing and insertion are leading to numerous new groups pursuing their own CARs across the world (especially in China).

It is now possible to insert several genes at a time into a T cell, which may lead to CARs with multiple protein targets. Ongoing research suggests it may also be possible to edit the T cell so that it has built-in defenses against the tumor microenvironment. CAR-T is also being tested in combination with checkpoint inhibitors and other immunotherapies.

Vaccines

The vaccines we were building even ten years ago had the right idea, but they were based on biology that was poorly understood and technology insufficient to executing these approaches effectively. Now the tech has caught up to the concept.[7] The buzzword now is "personalized cancer vaccines." Dan Chen explained it this way:

"We can take samples from the patient. We can sequence the whole genome really quickly, both the patient's genome and the tumor genome. The computer can take this totally ginormous data set and go boom boom boom, OK, here it is, here's your twenty top sequences. And we have a way to make a drug around those top sequences really fast. Will it work? We don't know. But the signs are really, really good."

Older vaccines are also now getting second looks following the discovery of checkpoint inhibitors, and our new understanding of how tumors manipulate, down-regulate, and suppress normal immune response. Researchers are currently reevaluating previously shelved cancer vacines (such as GVAX) in the new light of checkpoint inhabitation.

The Tumor Microenvironment and Other Targets

Tumors create a sort of microatmosphere, called the tumor microenvironment, which they poison with a smog of enzymes and immune inhibitors that choke out or shut down T cells. That environment surrounds the thousands of proteins a tumor expresses on its cell surfaces.

We are already familiar with some checkpoint inhibitors, but this is the tip of the iceberg. There are perhaps fifty potential targets to attack in this environment. Researchers are also exploring the arena of agonists, which stimulate (rather than inhibit) immune cells. Interesting and exciting research is ongoing on targets such as CD-27, CD40, GITR, ICOS, and others, although it's premature to speculate OX40, until more clinical data is available.[8] Cytokines are also the subject of a great deal of research and academic activity; in addition to the revisiting of the importance of IL-2, IL-15 is reported to be a reasonable candidate for future cancer immunotherapies. Additionally, there is renewed interest in the role played by other immune cells in T cell priming and activation, and how they might contribute to control of the immune-suppressive factors in the tumor microenvironment.

The role of macrophages, dendritic cells, natural killer cells, and others previously assumed to only be agents of innate immunity is a fast-moving research front, as are new findings about the role played by the microbiome of the gut in immune modulation, signal induction inhibitors (such as BRAF and MEK inhibitors), as well as microbiota alteration, activation of antigen presenting cells, targeting of cancer stem cells on tumor strata, and factors including nutrition, exercise, and even sunlight.

One of the takeaways from this list may be the simple fact that immunology is complicated and involves a lot of players. Basic scientific research is required to better understand these players and to understand them in regards to cancer. Immune response is a complex conversation. We have only started to learn how to listen.

APPENDIX A

Oncolytic Virus Therapy

An exciting and somewhat separate approach within immunotherapy uses viruses to selectively sicken and kill tumor cells without harming normal body cells. The result is essentially a disease
for your disease—a disease that makes only cancer sick. As of this
writing the only FDA approved version of this therapy, called
talimogene laherparepvec (brand name Imlygic), or T-Vec, uses a
genetically modified version of the herpes virus to infect melanoma
cancer cells. The melanoma is reprogrammed to create immune-
stimulating proteins and more of the cancer-infecting virus; in
time the melanoma cell bursts, spewing telltale tumor antigens
that alert the immune system to join the attack. This approach (in
combination) is showing greater success against some tumors than
checkpoint inhibitors alone and is being investigated as a means to
turn cold tumors (which for whatever reason repress or avoid the
attentions of the immune system) into hot tumors.

Biomarkers

Most cancer immunologists point to the problem of so many new
therapeutic options for patients who can ill afford the time or
resources of the wrong approach. Tests are needed, capable of cat-
egorizing both a patient's immune system and the specifics of their
cancer, in order to help clinicians of the future determine the most
effective therapy. Some clinicians and researchers are now calling
for an evaluation of a patient's "immunoscore" as an important early
step in treating cancer.

The Breakthrough, in Brief

Given the right conditions, the human immune system is capable of recognizing and killing cancer. And perhaps, eventually, an immune approach is the best way to get to a possible cure. And yet, for some reason, it hadn't worked. For years, cancer immunologists were racing to figure out why.

The immune system, like cancer, is a nimble, adaptable, and evolving system. Cancer had already proven its ability to rebound from the most direct attacks by drugs or radiation, a confounding, unique capacity now known as "escape." Even as a drug targeted cancer, that cancer mutated and evaded the attack. Whatever cells survived came roaring back, impervious to the old drugs. That mutational capacity defined cancer. But adaptability and mutation were also what defined immune response.

The immune system did a great job with most invaders to the bloodstream; it found sick cells and attacked and killed them. Cancer was a sick cell—a mutant cell in our own body that couldn't stop growing. So why didn't what happened with the common cold seem to happen with cancer? For decades, researchers had believed they were missing some pieces of the puzzle, the molecular keys that might allow the immune system to treat the diseases known as cancer in the same way that it treated other foreign pathogenic invaders, like viruses, bacteria, or even a splinter. Exactly why cancer seemed to receive a different immune response than other diseases, and exactly how it somehow evaded the complex web of traps and scouts, trackers and killers that patrolled the perimeter of our epidermis and floated invisibly in our bloodstreams, had been a subject of fierce debate. Most researchers believed that the immune system simply could not recognize cancer as a foreign (or "non-self") cell, because it was too similar to normal, healthy "self" cells.

A handful of stubborn cancer immunologists disagreed. They believed something about cancer allowed it to evade—and trick—the immune system's hunter and tracker cells. They were correct. Cancer uses these tricks to prevent its own destruction.

Only a few years earlier, that viewpoint was considered ridiculous by most cancer specialists, and perhaps hopeless, even by the few cancer immunologists still hanging on to the dream. But in 2011 some important new discoveries—breakthroughs in cancer research—finally identified some of the missing puzzle pieces that prevented the immune system from recognizing and attacking cancer. Much of this was good old-fashioned research that had nothing to do with cancer specifically.

Some of the mysteries of the immune system were finally teased out; the existence and role of the T cell as the serial-killing attacker of foreign cells was firmly established. The specific ignition switch for that immune response—a receptor on T cells that got "turned on," or activated, by recognizing the unique protein fingerprints (or "antigens") on sick or infected cells—had been identified, as was the mechanism of the amoeba-like dendritic cell, which was something of a frontline water boy of the immune system, presenting those antigens for the T cells to pick up and learn from. That communication gave a T cell its marching orders, like a wanted poster: it told the T cell what specific unique sick-cell surface proteins to look for, then sent the T cell off on a search-and-destroy mission. It was like a suspect's description being broadcast to police via an all-points bulletin.

The discovery of the receptor of the T cell (T cell receptor, or TCR) in 1984 and its subsequent cloning had helped finally pin down the means by which the T cell interacted with its pathogen target. The receptor of the killer T cell was a physical thing that fit the antigen it was supposed to target and kill like a lock to a key. It's through this lock and key, receptor and antigen interaction that the T cell is activated, and immune response against sick, or non-self, cells occurs.

But of course, because it's the human immune system, it couldn't be so simple. Researchers quickly realized that more than one key was required to initiate that immune response—something like how multiple keys are required to unlock a nuclear button or to open a safe deposit box. And for much the same reason.

The immune system is powerful and therefore dangerous. Proper triggering of immune response against pathogens is what keeps you healthy. But improper hair-trigger immune response against the self cells of your own body is autoimmune disease. This is a belt-and-suspenders approach to a life-or-death decision on a cellular level. If it wasn't safe, you'd be sorry.

The code was truly cracked with the discovery of that second signal required to activate the T cell. But that discovery had come with a surprise.

They'd been looking for a second signal, another "go" button that would act as a sort of gas pedal for that T cell and start the whole cascade of reactions that we call immune response, which results in killing the bad guys. But instead of a gas pedal, what reseachers discovered was a brake.

The brake, called CTLA-4, was useful for self cells preventing a T cell from an autoimmune attack. Allison discovered that cancer had hijacked that brake signal. The brake wasn't a key—it was a safety switch. CTLA-4 was a checkpoint. Cancer used it to shut down T cell activation before it ever began. By developing a drug (an antibody) that bound to and blocked the brake, they prevented the tumor cell from trying to use it. They kept cancer's foot off the brake of the immune system.

That breakthrough discovery inspired researchers to rethink and look harder for other checkpoints, and maybe other brakes. Blocking CTLA-4 worked, the way blocking the brake on a car prevented it from being pressed. But, to continue the car analogy, driving without brakes wasn't entirely safe, either. It worked, but the brake was a safety to control against autoimmunity.

For patients whose immune systems were not particularly responsive and who had tumors with obvious mutations that made them clear targets for an awakened immune system, there had been some remarkable results—tumors that melted away, terminal cancer that disappeared and never came back. For other patients, however, it was like driving in a car without brakes. Especially for patients with hair-trigger immune systems, CTLA-4 could be a ride from hell. And if those patients had cancers that were subtle and difficult for T cells to pick up on, that hell ride was too hard on the body, and not hard enough on the cancer. Like a fever that rages too high, it hurt faster than it helped.

But the proof of concept inspired researchers to consider other, more recently discovered cell receptors on the T cell. These, they hoped, would be more specific, awakening the immune response in a more intimate way, when the T cell was up close and personal to a tumor cell, and only in that up close environment.

Such a checkpoint inhibitor, if it existed, might have less severe side effects and better specifically targeted anticancer effects. And a potential second checkpoint inhibitor had been identified, another antigen on the T cell surface they called "PD-1." In some cancers, the tumors were discovered to have a complementary protein on their surfaces that fit into PD-1 like the other side of a handshake. The thing that fits a receptor is called a "ligand," so on the tumor side, they called it PD-Ligand 1, or PD-L1 for almost short. Tests in the dish and mouse models had led researchers to suspect that PD-1/PD-L1 was in fact a more precise and localized secret handshake between cells, which allowed cancer cells to convince T cells not to kill them. Normally, this was a handshake between killer T cells and body cells; cancer cells had successfully adopted this trick to stay alive. The hope was that if researchers could find a way to block that handshake, or checkpoint, they could block the trick, and the immune cell could kill cancer. Those "checkpoint inhibitor" drugs would be anti-PD-1, blocking the handshake on the T cell side, and anti-PD-L1, blocking it on the tumor side.

CTLA-4 cracked open the door; PD-21 blasted it wide. Suddenly years of failed experiments in cancer immunotherapy could be explained by the simple fact that they'd been trying to drive the immune system with the hand brake on. And for the first time, they suspected they might know how to turn it off.

They didn't believe it would work for all patients, or in all cancers. They didn't even know if it would be enough to make a difference. But the strong suspicion was that for some patients, simply removing the hand brake on the immune system, and allowing it to recognize the cancer as the non-self pathogen that it really was, might help make the other therapies they were using more effective. And, they suspected, for some patients, simply unleashing the immune system to do its job would be enough to destroy cancer.

This was the proving time, and an exciting period for immunologists who had spent careers looking for a missing piece to the immune puzzle. The first generation of checkpoint inhibitor drugs, the anti-CTLA-4 drugs, were already in the process of phase 2 trials. They were being tested in people—not just in phase 1 tests to see if they were safe, but now to see if they were effective. Despite some early hope, those trials were now having some significant problems. Two major pharmaceutical companies were testing their versions of these checkpoint inhibitors independently. The results had been discouraging enough that one of them had abandoned the trials, at an expense of millions of dollars and many years of work. The fate of the other was still uncertain, but the results thus far wouldn't pass FDA muster. The jury was still out on whether checkpoint inhibitors would just end up being another overhyped chapter in the history of immunotherapy, another approach that worked in mice and failed in humans, like the cancer vaccines.

Regardless, the new discovery of CTLA-4 had put other pieces of the immune puzzle in motion, including amped-up research and clinical trials on the other, newer checkpoint inhibitors. The stars among them are the anti-PD-1 drugs targeting the T cell side of the programmed cell death (PD) secret handshake, and anti-PD-L1, which blocked the tumor side.

These drugs would turn out to be game change for several types of cancer.

A Brief Anecdotal History of Disease, Civilization, and the Quest for Immunity

While we've only recently developed a reliable immune-based therapy against cancer, we have had immune system–based therapies against disease for centuries. The most familiar form of these immune system–based medicines is the vaccine. A vaccine is an agent intentionally introduced into a living body in order to stimulate a specific and direct protection against a specific disease. In its most basic form, this could be a crude introduction—say, a scratch—with the corpse of one of the dead pathogens. There's a lot of information on a bacterial corpse. You can think of them as hints and insights about the enemy you may one day face. And the immune system is a quick learner.

We owe the word *vaccine* to cows (it comes from the Latin *vacca*, for "cow"), and the vaccine itself to observations of the work of milkmaids.

Edward Jenner observed that persons who milked cows often developed a bovine-borne disease called cowpox, and those who did were also far less likely to catch its deadly human cousin, smallpox. In 1796 Jenner re-created this incidental inoculation. He used pus scraped from the blisters of a milkmaid named Sarah Nelmes; Nelmes's cowpox had been caught from a heifer named Blossom. Jenner then transferred the pus to the eight-year-old son of his gardener. His experiment inoculated the boy, and in turn granted the concept of intentionally engineered immunity scientific recognition and acceptance.

Jenner invented the modern vaccine—the ones you get at Walgreens to get immunity from this year's flu are essentially the same—and his breakthrough saved millions of lives. He was the first to use the scientific principle of borrowing one person's immune response (the pus) to a weaker cousin of a disease and weaponizing it for another person, making them immune to the disease itself. But even in the seventeenth century, the concept of immunity was nothing new. It had been familiar for so long that we took it as a sort of folk wisdom—common sense, even. People who had survived exposure to a disease usually weren't susceptible to that disease the next time it came through. That sort of thing is almost impossible not to notice.

The words in Latin are *immunitas* and *immunis*. Both referred to a legal concept of exception. In ancient Rome immunity was a legal pass from a citizen's usual responsibilities or duties, such as an obligatory period of military service, or taxes. With poetic license the first century AD Roman poet Lucan uses the word to describe the Psylli tribe of North Africa, whose members were famously said to be "immune" to snakebite. (See Arthur M. Silverstein, *A History of Immunology.*)

In fact, immunity from the terrible toll of poisons was a particularly well-subscribed field of study, being popular with those who had both the need to survive assasins as well as the gold to pay for such protection.

In her *Milestones in Immunology: A Historical Exploration*, Debra Jan Bibel cites the relatively more recent desire of kings, fearing death and succession by poisoning, to seek immunity from poison. In the first century we have a record, or perhaps a fable, of King Mithridates VI, whose kingdom of Pontus bordered the Black Sea. Mithridates attempted to give himself such immunity by taking a daily prescribed dose of the poison he assumed would be used to assassinate him. This assumes the state of fable because he reportedly succeeded, grew old, and wanted to end his life by poisoning himself—only to discover he was truly immune, and could not.

By the fourteenth century *immunity* had come to have implications of a special dominion of exemption from the toll of disease, bestowed by God. (This and the following come from Antoinette Stettler's *Gesnerus* review of "History of Concepts of Infection and Defense," as quoted by Silverstein in his *History of Immunology*.) *"Equibus Dei gratia ego immunis evasi,"* wrote Colle, referring to his escape from the plague epidemic.

Plagues and pestilence were a common feature of the ancient world. The plague that ravaged Athens in 430 BC killed an estimated 25 percent of the city's population. The historian Thucydides recorded the incident and made the observation that it was those Athenians who had been sickened and recovered who were best able to care for the dying: "They knew what it was from experience, and had now no fear for themselves; for the same man was never attacked twice—never at least fatally," he wrote. The phenomenon Thucydides is unwittingly describing is acquired immunity.

This was an early observation but one made repeatedly during pandemics; for example, one thousand years later, the historian Procopius described another plague, this one named for the emperor of Byzantium as the Plague of Justinian: "It left neither island nor cave nor mountain ridge which had human inhabitants; and if it had passed by any land, either not affecting the men there or touching them in indifferent fashion, still at a later time it came back; then those who dwelt round about this land, whom formerly it had afflicted most sorely, it did not touch at all" (Procopius, *The Persian War*, vol. 1, trans. H. B. Dewing [London: Heinemann, 1914]).

As a folk medical practice, inoculation had a long history but no scientific explanation. The Moors and the Pouls of the largely Muslim West African region that lies between Senegal and Gambia would stab a knife through the lungs of a cow that had died due to pleuropneumonia, then use it to make incisions in the hides of their healthy cattle. It was a de facto inoculation, introducing a bovine pneumonia to the immune system of healthy cows. Whether success was considered dependent on the specific knife or the specific person making the incision, or words incanted during the ritual or the design of the cuts is not known; the practice, reported in Western scientific journals in 1885 (e.g., *Compus Rendus de l'Académie des Sciences*), was already said to have an origin "lost in the obscurity of history."

As Silverstein wrote, the keen observer "could not help but notice that often those who by good fortune had survived the disease once might be 'exempt' from further involvement upon its return." This was the same phenomenon later to be described more scientifically by Jenner, who's credited with the great experiment that brought about the smallpox vaccine, and "intentionally" acquired immunity.

In 1714, two Greek-Italian physicians reported on some of these innoculation rituals as they applied to humans for the Royal Society of London, which was a sort of clearing house of officialdom for Western medicine. The disease in question was smallpox.

The first recorded epidemic of this disease farther west occurred in Arabia during the sixth century:

In 570, an Abyssinian (Ethiopian) army provided with war elephants, under the command of the Christian zealot Abraha Ashram, set forth from Yemen (then occupied by the Abyssinians) and attacked Mecca (now in Saudi Arabia) to destroy the Kaaba in that city. Kaaba was the sacred shrine of the Arabs, who were then heathen and kept their idols in it. According to Muslim tradition, this shrine was built by Abraham, the father of Isaac and Ishmael, whose descendants are the Jews and the Arabs, respectively. As stated in the Koran, the holy book of the Muslims, God sent flocks of birds which showered the attacking army with stones producing sores and pustules that spread like a pestilence among the troops. Consequently, the Abyssinian army was decimated, and Abraha died from the disease; thus, the Kaaba was saved from destruction. The year AD 570, which is also the birth year of Mohammad, the prophet of Islam, was designated by the Meccans as the year of the elephant. Medical historians have interpreted the above pestilence as an outbreak of smallpox which introduced this disease to Arabia from Africa (A. M. Behbehani, "The Smallpox Story: Life and Death of an Old Disease," *Microbiological Reviews*, 1983, 47:455–509).

Descriptions of smallpox appear in the most ancient Indian, Egyptian, and Chinese medical writing. Pharoah Ramses V appears

to have died of the disease in 1157 BC. Behbehani cites the transla-
tion of Arabic medical books into Latin by Constantinus Africa-
nus (1020 to 1087) as the origin of the name "variola" for the disease
described in 910 by the eminent Islamic physician Rhazes. For centuries it was considered an innocuous disease, but somewhere in
the tenth century it transformed into a more virulent strain, which
returned from the holy land with the crusaders in the following centuries. By the sixteenth century it had traveled on slave slips to the
West Indes, and from there to Central America and Mexico, where
it decimated populations and was at least partially responsible for
Hernán Cortés's ability to conquer the mighty Aztec empire with
only five hundred men and twenty-three cannons. In his wake the
disease continued its ravages, killing more than three million persons, while Cortés himself traveled to Cuba, the disease in tow. Five
years later it would travel across the isthmus to Peru, where it decimated the Inca and wiped out entire Amazonian tribes across South
America.

By this time it had also hopped the channel to Britain, and by
1562 it had infected Queen Elizabeth 1. The monarch survived the
disease, but it left her bald and facially disfigured. By the seventeenth century, outbreaks were deadly and regular occurrences. It
is estimated that at this time, the disease was responsible for four
hundred thousand deaths in Europe each year, and caused one-
third of the cases of blindness. Urban centers were hit especially
hard by contagious diseases, and the teeming streets of rapidly pop-
ulating London suffered disproportionately.

In a series of letters by Emanuele Timoni, a doctor affiliated
with the Royal Embassy in Constantinople, he and his colleague
Jacob Pylarini informed the esteemed scientific body of a folk

practice known as "buying the smallpox," which involved inoculating against smallpox, by collecting the hard, scab-like crusts that formed on the weeping pustules of someone who had been sickened, but not killed, by the disease—what the authors referred to as "favorable" cases of smallpox. These crusts would then be inserted directly into cuts in the skin of a smallpox innocent. Apparently the practice was unfamiliar to London society, but as Timoni and Pylarini witnessed, it was common practice and protection in the far eastern capital of Constantinople.

A British surgeon in Turkey described the practice as being performed by old women, who "scarred the wrists, legs and forehead of the patient, placed a fresh and kindly pock in each incision and bound it there for 8 to 10 days, after this time the patient was credibly informed. The patient would develop a mild case, recover, and thereafter be immune."

In fact, the practice was long known to rural communities in Western Europe, the Middle East, North and West Africa, and Asia, where writers have speculated that the practice may have begun. In China, it is described by Chinese author Wan Quan in his 1549 medical volume *Dou zhen xin fa*. Here, rough custom had been refined with some elegant touches; the smallpox scabs were ground into a powder and blown into the nose of the person to be inoculated, using a special silver straw (boys were inoculated through the left nostril, girls through the right). The inoculation was far from perfect, and sometimes served as an intentional infection. Using live smallpox was reported to kill as many as 2 percent of the participants and turned the rest into temporary contagious carriers of the disease. Still, that was considered a favorable trade-off for the 20–30 percent mortality rate of the disease itself.

Initial resistance to using such foreign techniques in London was worn down by the addition of courtly charm and title in the person of Lady Mary Wortley Montagu, a poet and travel writer known for a beauty marked by a fierceness of the eyes, and whose husband, Lord Edward Wortley Montagu, had been appointed ambassador to Constantinople in 1716. Lady Mary traveled with him and observed the Turkish custom of variolation.

She had already survived the disease herself, which had scarred her face and caused her to lose her eyelashes—one potential source of their noted intensity. Her brother had been less fortunate. She was impressed with the local custom, enough to insist that the embassy surgeon inoculate their five-year-old son, Edward Jr., while her husband was away on official business at the Grand Vizier's camp in Sophia. The embassy chaplain protested that the procedure was "unchristian" and could work only on "infidels," but Lady Mary was persistent. Dr. Charles Maitland inoculated one of the boy's arms with a lancet, an "old Greek woman" inoculated the other "with an old rusty needle," and both presumably used the method Lady Mary described in her letters, with the pus of an eleven-year-old that had been extracted into a small glass bottle and kept at a proper temperature in the armpit of the physician. That boy's apparent immunity made Lady Mary a enthusiastic booster for the "Turkish method," which she called "ingrafting." When she returned to London in 1721 she had the same embassy surgeon, Charles Maitland, repeat the procedure on her then four-year-old daughter. The technique was already customary in the countryside for nobody knows how long, but this was the first time it was performed by a medical professional, and certainly the first time it was done in view of royal court physicians. The little girl's pale, thin arm was exposed, slight incisions were made, and as the blood ran and the brave girl allowed a stranger's scabs to be stuffed against the wound, Sir Hans Sloane watched and considered.

Sir Sloane was an eminent physician, both president of the Royal Society and personal physician to the king. Lady Mary had been a vocal advocate for the method ever since her first letter back from Constantinople. She was a lady of social status, articulate and worldly and beloved by London society—but she was neither a physician nor a man. Sir Sloane, of course, carried both of these contemporary qualifications. His opinion, and that of the community of physicians at large, was that variolation was a dangerous procedure. But soon, news of the little girl's successful recovery and immunity triangulated with Sloane's eyewitness bona fides and Lady Mary's example.

In the summer of 1721, London was in the midst of a smallpox epidemic. Among those hoping to escape its ravages were the royal family. Five reigning European monarchs (Joseph I of Germany, Peter II of Russia, Louis XV of France, William II of Orange, and the last elector of Bavaria) would succumb to this disease during the eighteenth century. The princess of Wales, Caroline of Ansbach, very much hoped to spare her own children this fate. She was familiar with Lady Mary through social circles, and was known as a bright and scientifically minded royal, interested in the advances of her day. (The court flatterer Voltaire referred to the princess as "a philosopher on the throne.") Further convinced by the royal court physician, she and her husband (the future George II) agreed to sponsor a sort of clinical trial, of the sort that would pass exactly zero twentieth-century ethics boards.

By late July 1721 arrangements were made with the officials of London's notorious Newgate Prison; six prisoners were to be selected with the help of the royal physician and the apothecary from the ranks of those condemned to hang for their crimes. These would be the human guinea pigs. In exchange, they would be granted their freedom—immunity for immunity. It wasn't clear, however, whether they'd be alive to enjoy it.

On August 9, Dr. Maitland repeated the procedure on the prisoners, three men and three women, aged nineteen to thirty-six years of age. They were variolated on their arms and right legs as a group of twenty-five physicians, surgeons, and apothecaries watched. Five of the six developed smallpox symptoms by August 13; the sixth turned out to have already had the disease, and was already immune. All made a complete recovery and were granted their freedom, as promised.

But to test their immunity, the nineteen-year-old female prisoner was hired as a temporary nurse by the royal physician and transported to the town of Herford, which was suffering under the ravages of a particularly intense smallpox outbreak.

The girl worked as a nurse to a smallpox patient during the day, and at night was boarded in the same bed with another smallpox patient—a ten-year-old boy. After six weeks of this work, the young woman still showed no sign of the disease.

Newspapers covered the story of the experiments sponsored by the royal couple. They were generally favorable. (At this time a physician also variolated another female prisoner, this time through the Chinese method of powdered crusts blown into the nasal cavity. The newspapers at the time were highly critical of this experiment, because apparently the woman has been asleep when it was done.)

Soon volunteers began asking for the same treatment. Thus proven, the royal daughters—eleven-year-old Amelia and nine-year-old Caroline—were variolated on April 17, 1722.

The procedure received the sort of attention that attends all affairs of royal children great and small, but it was hardly a cure for the disease, only a better roll of the dice. Maitland, while concluding the experiments in Hertford, had privately variolated a number of children from private households; one had become ill and spread the smallpox to six household servants, one of whom died.

This pattern was repeated in other households, where servants were exposed to inoculated children and succumbed to the disease. Others who lined up to receive the fruits of what would become known as "the Royal Experiment," such as the child of the Earl of Sunderland, did not recover from the illness and died days later.

Priests railed against the unnatural regimen from the pulpit, telling their flocks that "the dangerous and sinful practice of inoculation" was diabolical, promoted vice, and "usurp[ed] an authority founded neither in the laws of nature or religion." The London surgeon Legard Sparham published a pamphlet against inoculating, and articulated his reasons against inserting diseases into healing wounds, calling it "bartering health for diseases." (As we have seen, this "deal" had echoes in late-nineteenth-century New York City and in the foundational observations of cancer immunotherapy.)

But the variolation treatment received a more broad endorsement from the London Royal Academy, more so after its secretary and mathematician examined the results statistically. He found that death from variolation between 1723 and 1727 occurred in between 1 out of 48 to 1 out of 60 cases, while death from natural smallpox occurred in 1 out of 6. Royal opinion had been vindicated.

Inoculation would be the law in Britain. That sensibility didn't necessarily translate to the rustics in the colonies, however—a fact that almost decided the War of Independence and ended America's revolution.

~

We don't know if Onesimus was his given name, that record is lost. It's believed that he came from the Fezzan region of southwestern Libya, a land of rocks and high dunes surrounding the oasis capital of Murzuq, though it's impossible to be sure. (At the time Murzuq was a thriving hub for both pilgrims and the slave trade that fed captives from Chad and the Central African Republic.)

What is certain is that as a young man, Onesimus was inoculated against smallpox in the Ottoman fashion, and the variolation left him with a telltale scar, and that at some point around 1718, Onesimus was kidnapped by slavers and shipped in chains to the American colonies to be sold at auction.

Since the seventeenth century, the center of the American slave trade was the port of Boston. Here Onesimus was purchased at auction by a man of God and science named Cotton Mather. Cotton Mather seems an especially curious figure—a widely read champion of learning best remembered for his involvement in the Salem witchcraft trials, a man of strict religious character who owned other human beings. None of that marked him as extraordinary in eighteenth-century Boston. What did make Mather unusual was that he was literate, well read, and both observant and curious about the world around him. Now he was curious about the variolation marks on Onesimus's arm.

Onesimus had unwillingly carried the technology of inoculation from Muslim northern Africa to the primitive American colonies. Mather was sharp enough to be curious about the practice, and did not understand why it wasn't in the colonies as well.

In June 1721, the disease that raged in London the summer before arrived at the American colony via the HMS *Seahorse*, recently of the West Indies. Soon the disease showed all the hallmarks of an epidemic. It would be devastating to the small capital—a city in name only, laid along the paths worn by cattle and sheep. In matters of contagion, Cotton Mather was one of the few people in Massachusetts qualified to give advice.

This was a small, rough world, and Mather's learned religiosity was leavened by an equally giant intellect that towered over most of his largely illiterate colonial peers. The few men who did read knew each other, and borrowed one another's books. (Mather lent and borrowed volumes with Benjamin Franklin, then a precocious young apprentice at a printer's shop near Mather's home; he also had his own pamphlets printed at Franklin's shop. Mather's business helped Franklin establish himself in his own print shop; the lending and borrowing practice among the small community of readers led Franklin to initiate the colony's first lending library.)

Mather was not a physician, but he read their journals when he could and was up-to-date on the most recent advances. (More so than most physicians, though this is not surprising; in all the colonies at that time there was only one working physician who held a medical degree—his book-lending acquaintance, Dr. William Douglass, formerly of Edinburgh University.)

Douglass subscribed to the latest medical journals from abroad. Mather borrowed them and found Timoni's published letter to the Royal Academy in London regarding the practices of smallpox variolation as witnessed in Constantinople. The method Onesimus described was now mirrored in a medical journal, and validated by the Royal Academy of mother England, a holy trinity of conviction for a man like Mather.

Mather's was a radical view, intellectually, not just for 1724 Boston but for the larger scientific community, and even more radical when Mather attempted to act on it. The colony's only properly degreed physician violently opposed Mather's inoculation attempts. During the year 1721 Mather spent a small part of his energies trying to press variolation technique onto the medical men of Boston. He convinced only one, a medically inclined stone cutter named Zabdeil Boylston. Boylston preformed the procedure on his son, his slave, and his slave's son. All three survived and were safely inoculated, but the intellectual backlash was pronounced.

Boylston was attacked in the papers, and then physically by mobs in the street. Mather, undeterred by Boylston's beating, then inoculated his own son in the same way. The procedure sickened the boy and almost killed him, which only made his fellow colonists more fearful and angry. Mather was seen as spreading disease and risking pandemic. Every smallpox victim was a potential grenade of illness in the small rural community, and as retort, at three o'clock that night an angry antivaxxer lobbed a real grenade through Mather's window and into the house where Mather's son and another minister, recovering from his own smallpox variolation, were then recuperating. The grenade survived—the lit fuse apparently became detached as it crashed through the window— and was found with an anti-inoculation note attached.

Boylston would report that by 1722 he had inoculated 242 people in the Boston area, of whom six died—a 2.5 percent mortality rate. That could be compared to a reported 15 percent mortality rate of cases of natural smallpox in the Boston area, 849 deaths among 5,889 cases of naturally occurring disease. Variolation involved treating healthy people with a deadly disease; it worked sometimes, it was true, but it did so through a mechanism that was beyond the best scientific minds of the time. Where man interfered in the natural order, whatever magic resulted might be of demonic design.

The truth was wonder beyond the imagining of any contemporary watchmaker or apothecary.

Variolation would eventually gain greater appreciation in America, but it still lagged behind Britain. Several American states passed laws against it; some colonial cities declared themselves to be antivariolation zones and became sanctuary cities for antivaxxers.

George Washington believed in the effectiveness of the technique, however, and had his troops inoculated before the siege of Boston. But these inoculations were risky—during the infectious stage, the treatment could trigger an epidemic—and grudgingly, Washington stopped the program. Historians now believe that as a result, smallpox ravaged the colonial army in a way that it had not affected the variolated British forces, beneficiaries as they were of the monarch's royal experiment.

It's been suggested by some historians that it was smallpox and the antivariolation cities of the north that saved Canada for Britain.

Further Reading

Abbas, Abul K., Andrew H. Lichtman, and Shiv Pillai. *Cellular and Molecular Immunology* (eighth edition). Philadelphia: Elsevier Inc., 2015.

Bibel, Debra Jan. *Milestones in Immunology: A Historical Exploration*. Madison, WI: Science Tech Publishers, 1988.

Butterfield, Lisa, ed. *Cancer Immunotherapy Principles and Practice*. New York: Demos Medical Publishing, 2017.

Canavan, Neil (The Trout Group LLC). *A Cure Within*. Cold Spring Harbor, NY: Cold Spring Harbor Laboratory Press, 2018.

Clark, William. *A War Within: The Double-Edged Sword of Immunity*. New York: Oxford University Press, 1995.

Hall, Stephen S. *A Commotion in the Blood*. New York: Henry Holt and Company, Inc., 1997.

Mukherjee, Siddhartha. *The Emperor of All Maladies*. New York: Scribner, 2010.

Rosenberg, Steven A., and John M. Barry. *The Transformed Cell*. New York: G. P. Putnam's Sons, 1992.

Silverstein, Arthur M. *A History of Immunology* (second edition). London: Academic Press, 2009.

Thomas, Lewis. *Lives of a Cell: Notes from a Biology Watcher*. New York: Viking Press, 1974.

Wilson, Edward O. *Consilience: The Unity of Knowledge*. New York: Knopf, 1998.

Notes

Introduction

1. Amazingly, it was less than a year after the 1895 discovery of mysterious, or "X," rays by German physicist Wilhem Röntgen that high-energy electromagnetic radiation machines became a medical technology. Homeopathic physician Dr. Emil Grubbe had one at the Hahnemann Medical College of Chicago, to treat carcinoma, but those early machines may have done more harm than good; Dr. Grubbe himself endured more than ninety operations for multiple cancers. Grubbe was an early adopter. It was Marie Curie's discovery of radioactive elements that greatly broadened the use of radiation as a cancer therapy. Both Röntgen and Curie received Nobel Prizes for their work. Titus C. Evans, "Review of *X-Ray Treatment—Its Origin, Birth, and Early History* by Emil H. Grubbe," *Quarterly Review of Biology*, 1951, 26:223.

2. These numbers are meant as a very broad illustration of concept and should in no way be confused with any statistical probability or scientific certainty that a rogue cell is not recognized by the immune system or will grow out to become something we would recognize as clinical cancer. The larger point is that the immune system is generally extremely successful at recognizing non-self and that random, infinite rolls of the dice render even the most exceptionally improbable outcomes inevitable. Factors like viral infection or certain chromosomal defects make those improbable outcomes less so.

3. Called pembrolizumab, it is a humanized monoclonal antibody that binds to and blocks the T cell's PD-1 receptor. It is manufactured and sold by Merck under the brand name Keytruda.

4. Mr. Carter had metastatic melanoma, which had spread to his liver and brain, for which he underwent surgery and received radiation therapy, in addition to his immunotherapy.

5. The first to suggest this analogy to me was Dr. Daniel Chen.

One: Patient 101006 JDS

1. He made enough to get his own season tickets—Yankees, twenty-two years—not that he used all the tickets after he and his wife, an executive at the time with Interscope Records, moved to California.
2. He would have been cool with accounting regardless, but no doubt, representing rock-and-roll acts was the coolest version of accounting.
3. "Yesterday" is the most covered song of all time, according to the *Guinness Book of World Records*. It's not difficult to believe it's made royalties comparable to those of "Yellow Ribbon."
4. Jeff's diagnosis occurred in 2011. While it's possible that he would have responded to the T cell side of this ligand interaction, the future of PD-1 was in that moment in a sort of limbo in terms of patient availability, as we'll see in later chapters, and would not be approved by the FDA until 2014, years after Jeff Schwartz began PD-L1 trials and significantly later than he was expected to survive. Further, that initial PD-1 approval was only for metastatic melanoma. Approvals for other indications have followed and are continuing.
5. Sutent, which targets a tumor's ability to eat and grow, is not, technically, "chemotherapy." Though Schwartz was not privy to the identity of the drug he'd be testing, it was for the anti-PD-L1 checkpoint inhibitor atezolizumab, which would be marketed under the trade name Tecentriq. See Appendix A.
6. Avastin (bevacizumab) has since been approved as part of a combination therapy for several different cancer types, including metastatic renal cancer, when used with interferon alpha.
7. Brian Irving, Yan Wu, Ira Mellman, and Julia Kim.
8. The Chens have three children: a daughter, Isabelle; and two sons, Cameron and Noah.

Two: A Simple Idea

1. A privately printed monograph, "Creating Two Preeminent Institutions: The Legacy of Bessie Dashiell," details aspects of the relationship beween Dashiell and John D. Rockefeller Jr., and links it to Rockefeller's subsequent foundational support of cancer research institutions Rockefeller University and Memorial Sloan Kettering Cancer Center. The slim volume was printed in very limited quantities in 1978 by the Vermont-based Woodstock Foundation (also Rockefeller supported), with a copy held in the collection of the Cancer Research Institute along with the Coley archive. Matt Tontonez, CRI's

excellent science writer (now with Memorial Sloan Kettering Cancer Center), generously guided me to this and other sources used throughout the chapter.

Additional information comes from the indispensable resource of William B. Coley's personal papers, collected by and significantly added to by his daughter, Helen Coley Nauts. Originally kept at CRI, they were donated upon. Nauts's death in 2001 to the Yale University Library (Helen Coley Nauts papers MS 1785), where they are now being catalogued. That collection consists of patient files, correspondence, writings, subject files, and other materials documenting the careers of Helen and her father, as well as extensive material regarding his toxins. This collection fills 119 boxes, 117.5 linear feet of material.

2. This Pullman detail, found in several journal articles (e.g., David B. Levine, "Gibney as Surgeon-in-Chief: The Earlier Years, 1887–1900," *HSS Journal: The Musculoskeletal Journal of Hospital for Special Surgery*, 2006, 2:95–101), surely sources from a personal interview with Coley's daughter, Helen Coley Nauts, by the author Stephen S. Hall, whose 1997 volume on the history of immunology (*A Commotion in the Blood: Life, Death, and the Immune System* [New York: Henry Holt]) deserves particular acknowledgment as an invaluable resource and a much-recommended read. Mr. Hall was also generous enough to (briefly) switch roles and submit to an interview, for which this author is grateful.

3. New York Hospital was the country's third oldest, established by a 1771 royal charter by England's King George III, for the "reception of such patients as require medical treatment, chirurgical management and maniacs." By the time Coley interned there it had outgrown its original building on Broadway between what is now Worth and Duane streets and had moved to a new building between Fifth and Sixth Avenues and West Fifteenth and Sixteenth Streets.

Plenty of fascinating detail can be found in an account of a physician's address before the New York Hospital alumni board: "Old New York Hospital; Its Interesting History Retraced by Dr. D. B. St. John Roosa. Episode of the Doctors Mob. The Aftermath of a Fourth of July Celebration. Forty Years Ago—Surgery Then and Now," *New York Times*, February 11, 1900.

4. Coley's medical education had been a late decision, made only after he decided against studying law and had spent two years as a schoolmaster teaching the classics in Oregon. He'd gotten into Harvard Medical School's three-year program as a second-year, owing to the experience of making horseback house calls with his physician uncle in rural southern Connecticut. He had been lucky enough, in his first year of residency, to be given a summer position replacing an

absentee doctor at New York Hospital. The direct observation of human illness put the five-foot-eight young man a head above his peers in applying for positions, and when he landed back at New York Hospital as an intern, he found himself mentored by two of the most famous and influential surgeons in the country, Robert F. Weir and William T. Bull. His later appointments would include the Hospital for the Ruptured and Crippled, now the Hospital for Special Surgery.

5. As a result of the improvements of Lister and others, a surgeon's art could be practiced with lessened probability of infection, which had plagued such procedures for thousands of years.

6. The date of the examination was October 1, 1890.

7. Rudolf Virchow especially had advanced the pathology of cancer as examined under the microscope, allowing it to be more systematically diagnosed.

8. Details of Dashiell's case can be found in William B. Coley, "Contribution to the Knowledge of Sarcoma," *Annals of Surgery*, 1891, 14:199–220.

9. W. B. Coley, "The Treatment of Malignant Tumors by Repeated Inoculations of Erysipelas: With a Report of Ten Original Cases," *American Journal of the Medical Sciences*, 1893, 105:487–511.

10. From Coley's collected writings we learn that Stein's sarcoma had first presented itself in 1880 as a small spot on his cheek, which by the following year had grown to a mass that required surgery. It came quickly back and was again removed surgically the following year. When Stein arrived at New York Hospital two years later, the mass had returned, and it reportedly looked something like a small bunch of grapes. This was the mass Dr. Bull operated on, and again in 1884—eventually creating the abscessed wound, which would not close with skin grafts and which finally became infected. Coley, "The Treatment of Malignant Tumors by Repeated Inoculations of Erysipelas"; William B. Coley, "A Preliminary Note on the Treatment of Inoperable Sarcoma by the Toxic Products of Erysipelas," *Post-Graduate*, 1893, 8:278–286; W. B. Coley, "The Treatment of Inoperable Sarcoma by Bacterial Toxins (The Mixed Toxins of the Streptococcus of Erysipelas and the Bacillus Prodigiosus)," *Practitioner*, 1909, 83:589–613; the archives of Cancer Research Institute; and other sources.

11. The German surgeon Friedrich Fehleisen established the association between the bacterium and the infection, as well as the first description of its appearance through a microscope. Friedrich Fehleisen, *Die Aetiologie des Erysipels* (Berlin: Theodor Fischer, 1883).

12. Until the discovery of antibiotics several decades later, there was no treatment for this infection.
13. Not just erysipelas but several other diseases, including ergotism and herpes zoster (shingles), had been referred to as St. Anthony's Fire, after the Christian saint to whom those afflicted would appeal for healing. Around 1100, the Roman Catholic Order of St. Anthony was formed in France to care for those with the ailment.
14. Also that year, two different scientists, one Swedish, and one a US geologist, had independently suggested that increased human production of CO_2 might result in global warming.
15. Emphasis mine. This 1909 reference and several others come from Coley's papers at the Yale University Library (Helen Coley Nauts papers MS 1785). This reference was also recounted in Stephen Hall's *A Commotion in the Blood*.
16. Anton Pavlovich Chekhov, *Letters of Anton Chekhov to His Family and Friends, with Biographical Sketch*, trans. Constance Garnett (New York: Macmillan, 1920).
17. Especially following fevers, infections, or both. A. Deidier, *Dissertation Médicinal et Chirurgical sur les Tumeurs* (Paris, 1725); U. Hobohm, "Fever and Cancer in Perspective," *Cancer Immunology, Immunotherapy*, 2001, 50:391–396; W. Busch, "Aus der Sitzung der Medicinischen Section vom 13 November 1867," *Berliner Klinische Wochenschrift*, 1868, 5:137; P. Bruns, "Die Heilwirkung des Erysipelas auf Geschwülste," *Beiträge zur Klinische Chirurgie*, 1888, 3:443–446.
18. Arthur M. Silverstein, *A History of Immunology*, 2nd ed. (Boston: Academic Press / Elsevier, 2009).
19. William Boyd, "The Meaning of Spontaneous Regression," *Journal of the Canadian Association of Radiologists*, 1957, 8:63; H. C. Nauts, "The Beneficial Effects of Bacterial Infections in Host Resistance to Cancer: End Results in 449 Cases," Monograph 8 (New York: Cancer Research Institute, 1990).
20. Ilana Löwy, "Experimental Systems and Clinical Practices: Tumor Immunology and Cancer Immunotherapy, 1895–1980," *Journal of the History of Biology*, 1994, 27:403–435.

In a similar experiment undertaken in 1868, a German scientist named W. Busch had intentionally infected a sarcoma patient with the erysipelas bacterium, *Streptococcus pyogenes*. After surgery, he had the patient recover in a hospital bed notorious for infecting every patient to inhabit it. This case was no exception, except that, as Busch reported, the resulting infection shrank the patient's tumors. Whether it was also responsible for her death nine days later was not specified, but that death did not take away from his larger point: induced infection seemed to have an effect on cancer when nothing else did. If they could control the infection, perhaps they might create a cure.

21. Coley, "The Treatment of Malignant Tumors by Repeated Inoculations of Erysipelas"; W. B. Coley, "The Treatment of Inoperable Sarcoma by Bacterial Toxins," *Proceedings of the Royal Society of Medicine*, 1910, 3(Surg Sect):1–48.

22. Zola case notes from Coley's collected papers, Cancer Research Institute archives.

23. Zola lived on Manhattan's Lower East Side, and his own niece risked her life to serve as his nurse. For her pains, she'd contract an erysipelas infection as well.

24. William B. Coley, "Further Observations upon the Treatment of Malignant Tumors with the Toxins of Erysipelas and Bacillus Prodigiosus, with a Report of 160 Cases," *Johns Hopkins Hospital Bulletin*, 1896, 7:157–162.

25. S. A. Hoption Cann, J. P. van Netten, and C. van Netten, "Dr. William Coley and Tumour Regression: A Place in History or in the Future," *Postgraduate Medical Journal*, 2003, 79:672–680; Coley, "The Treatment of Malignant Tumors by Repeated Inoculations of Erysipelas."

Possibly Coley's tepid language serves to cover for multiple scientific sins, including a failure to reproduce the remission experienced by Fred Stein, one Coley had only seen as briefly described anecdotally in Stein's handwritten medical report. Coley believed that the problem was in his patient selection for the experiments—their sarcomas were too far advanced; otherwise, he wrote, "It would not have been too much to have looked for a permanent cure." But here, Coley's vague and subjective language suggests that he wasn't seeing the results expected.

For all his insight, Coley was guilty of medicine's original and cardinal sin of putting ego before evidence. But he was onto something and he knew it, and he would soon prove willing to tie himself to the mast once that course was set.

Perhaps this is the countervailing genius of ego: persistence in the face of what's taken to be common sense. This was something I heard time and again while interviewing researchers. In science as everywhere, overconfidence in one's own ideas is hubris, but confidence, or at least a conviction in logic and empiricism even when the data point to unpopular or impolitic conclusions is essential to the seeker of truth. The real breakthroughs come from those who are tenacious, undeterred, and undiscouraged early in an experiment. They have courage and conviction. But in the end, they must also do good science, make nonsubjective observations, and present good data. Coley didn't do that. But his original observation had been astute, accurate, and important. He'd seen a spontaneous cancer remission and recognized it as science rather than miracle. And he doggedly pursued the experiments that might make use of that science.

26. This was the age of the bacteriologist, and even those who would not have regarded themselves thusly dabbled across the invisible and wholy arbitrary line separating the study of microscopy from that of the animals visible only through the microscope's lens. The interest in "disease" vs. the interest in bacteria recognized the relevance of toxins to both patient and pathogen. The chemistry of those toxins, and the responses of factors in the blood against those toxins follow from this study. Some of the more well-known names of this group include Louis Pasteur, Emil von Behring, Élie Metchnikoff, Paul Ehrlich, and Robert Koch. Their variously conflicting and complimentary approaches toward bacteria as the agents of disease gave birth to what would become the field of immunology, and what was then called seritology—the study of the fluid portion of blood, after the red and white blood cells had been strained out through the microscopic pores of a porcelain filter. Immune response was still a black box mystery, but it was known that bacteria were responsible for some diseases, and vaccination against those bacteria was possible. And all of that was presumed to happen in this colorless fluid in the blood.

27. B. Wiemann and C. O. Starnes, "Coley's Toxins, Tumor Necrosis Factor and Cancer Research: A Historical Perspective," *Pharmacology and Therapeutics*, 1994, 64:529–564.

28. Hall, *Commotion*, p. 57.

29. New York Cancer Hospital had been founded by the city's wealthy following news of the serial-cigar-smoking president Ulysses S. Grant's diagnosis of throat cancer. It was the first cancer-specific hospital in the United States, the second in the world. Its original mission was primarily to care for the dying in comfort and relative luxury. The wards were designed to accommodate the most modern, medically hygienic requirements for airflow and personal patient ventilation, garnering praise from a *New York Times* reviewer, but ultimately had a little too much style to be practical; the round rooms, designed in order to keep filth and germs from accumulating in the corners, did not give themselves to practical partition, and the gothic French chateau–styled turrets soon became as decrepit as the royal manors of old Europe they were designed to invoke. They were eventually abandoned for more practical quarters and slated for demolition but are now preserved as landmark multi-million-dollar condominiums with fabulous Central Park views. Christopher Gray, "Streetscapes/ Central Park West Between 105th and 106th Streets; In the 1880's, the Nation's

First Cancer Hospital," *New York Times*, December 28, 2003; New York City Landmarks Preservation Commission, Andrew S. Dolkart, and Matthew A. Postal, *Guide to New York City Landmarks*, 4th ed., ed. Matthew A. Postal (New York: John Wiley & Sons, 2009).

30. On April 21, 1892, Coley began injections on a forty-year-old man with an inoperable tumor in his back. It had been diagnosed as sarcoma and spread to his groin. After four weeks of steady injections a high fever was realized, and the patient began to respond in a manner similar to Zola's. Coley described this reaction with the enthusiasm of a nature writer witnessing his first sunset at Grand Canyon: "From the beginning of the attack the change that took place in the tumor was nothing short of marvelous. It lost its luster and color and had shrunk visibly in size within twenty-four hours. Several sinuses formed the second day and discharged necrosed tumor tissues. A few days later the tumor of the groin, which was about the size of a goose egg and very hard when the inoculations were begun, broke down and discharged a large amount of tumor tissue. Three weeks from the date of the attack of erysipelas both tumors had entirely disappeared." But Coley's excitement would not last. The man's tumors continued to grow back, and he continued to receive injections and surgeries before finally succumbing to a cancer in his abdomen three and a half years after his treatment had begun.

31. The subject of fever deserves its own chapter. The widespread use of antibiotics has saved countless lives from the ravages of infection, and they are routinely employed in postsurgical settings, where symptoms of infection, such as fever, are routinely suppressed. It may be worth considering whether this aspect of immune response is more than a symptom and side effect, and consider it a therapeutic aspect, with overlooked benefit. Fever has been shown to correspond to increased biochemical reaction rates and leukocyte proliferation, maturation, and activation. Fever also has been reported to have a palliative effect on pain. This paragraph is not a substitute for a more substantive examination of the topic, but a bookmark for further examination as to whether such a metabolically expensive physiological reaction triggered by the immune system would be conserved across species unless it had some benefit to survival.

32. Friedrich Fehleisen, the preeminent German physician using those same erysipelas injections at the Wurzburg clinic, had given up on the bacteria and been forced to resign his prestigious post due to patient deaths; this information appears in *A Commotion in the Blood* and was relayed to Stephen Hall by Otto Westphal.

33. Coley, "The Treatment of Malignant Tumors by Repeated Inoculations of Erysipelas."

34. The biological term is *attenuated*.
35. Experiments by physician H. Roger of the Pasteur Institute, reported in a mellifluous French that renders bacteria elegant. H. Roger, "Contribution à l'étude expérimentale du streptocoque de l'érysipèle," *Revue de Médecine*, 1892, 12:929–956.
36. *S. pyogenes* and *Serratia marcescens*. William B. Coley, "Treatment of Inoperable Malignant Tumors with the Toxines of Erysipelas and the Bacillus Prodigiosus," *Transactions of the American Surgical Association*, 1894, 12:183–212.
37. Though Coley didn't have the means to understand the immunologic and biochemical mechanisms of action of his toxin's reported therapeutic effects, more recent experiments have demonstrated that the erysipelas bacterium had nothing to do with it. However, in the 1970s, experiments by legendary cancer immunotherapist Lloyd Old at MSKCC and others showed that the other bacteria Coley had utilized, *B. prodigiosus*, did create endotoxins that stimulated the immune system's macrophages into producing powerful immune system messengers, cytokines including interferon (IF), interleukin (IL), and tumor necrosis factor (TNF). Boyce Rensberger, "Century-Old Cancer Treatment Reexamined," *Washington Post*, September 18, 1985.

    How these would have interacted with the tumor is another mystery, addressed in B. Wiemann and C. O. Starnes, "Coley's Toxins, Tumor Necrosis Factor and Cancer Research: A Historical Perspective," *Pharmacology and Therapeutics*, 1994, 64:529–564. Old and others (as we will see in later chapters) would continue to probe the mysteries of these molecules, and test them against a host of diseases, including cancer. Some were demonstrated to be powerful tools against tumors in animal models, and they were quickly, if prematurely, hailed on the covers of *Newsweek* and *Time* as the possible magic bullet cures for cancer; in humans they would be found to have powerful but uneven results— potent, sometimes curative, sometimes toxic, and generally poorly understood.

    Those cytokines are covered in more detail in later chapters; some are now being reexamined in the light of the newest cancer breakthroughs, as important pieces of combination therapies.
38. William B. Coley, "The Treatment of Inoperable Malignant Tumors with the Toxins of Erysipelas and Bacillus Prodigosus," *Medical Record*, 1895, 47:65–70.
39. Coley's work met a great deal of resistance, and many of the leaders in the sarcoma and general medical world actively worked to discredit his work, going so far as to suggest that his remissions were invaid because his diagnoses had been wrong, and his patients never had cancer to begin with. As usual, the takedowns were heard more clearly than the retractions. In 1934, the editorial board

of the *Journal of the American Medical Association*, which had declared Coley's lifework invalid, changed its mind:

> It appears, that undoubtedly the combined toxins of erysipelas and pro-
> digiosus may sometimes play a significant role in preventing or retarding
> malignant recurrence or metastases; occasionally they may be curative in
> hopelessly inoperable neoplasms;... The Council has, for these reasons,
> retained Erysipelas and Prodigiosus Toxins-Coley in New and Nonoffi-
> cial Remedies, with a view to facilitating further studies with the product.

Excerpted from "Council on Pharmacy and Chemistry: Erysipelas and Prodigiosus Toxins (Coley)," editorial, *Journal of the American Medical Association*, 1934, 103:1067–1069.

40. B. J. Johnston and E. T. Novales, "Clinical Effect of Coley's Toxin. II. A Seven-Year Study," *Cancer Chemotherapy Reports*, 1962, 21:43–68.

41. In 1992, the British journal *Nature* published a study by Charlie Starnes, a molecular immunologist who took a fresh look at William Coley's data on inoperable sarcoma cancers that had received no treatment aside from Coley's Toxins. What he found was, essentially, patient response rates far better than what anyone else had achieved to date in trying to treat these cancers by other means.

Some 10 percent of Coley's patients saw remissions of at least twenty years, many longer. Set against the baseline of otherwise hopeless cases and 100 percent remission, use of the toxins showed promise worthy of investigation.

By the standards of most modern cancer studies, in which *remission* refers to a patient showing no evidence of disease (NED) for at least five years following therapy, the results were even more striking; seventy-three of Coley's 154 sarcoma and lymphosarcoma patients—47 percent—were NED five years after treatment (Charlie O. Starnes, "Coley's Toxins in Perspective," *Nature*, 1992, 357:11–12).

It had taken nearly one hundred years before basic scientific research would catch up to what pioneering immunologist Lloyd Old called "the Coley Phenomenon"—the Tantalus-like purgatory of having perceived a mechanism in nature that might save millions of lives while lacking the tools to prove or use it.

"The cellular and molecular language of inflammation and immunity had to be understood," Old said, "before the forces that Coley unleashed could be predictably translated into tumor cell destruction."

42. Stephen Hall, in *A Commotion in the Blood*, quotes comments made by Nicholas Senn of Rush Medical College to his colleagues at the American Medical Association's 1895 annual meeting: "The treatment of inoperable sarcoma and carcinoma with the mixed toxins, as advised and practiced by Coley, has been given a fair trial

in the surgical clinic of Rush Medical College, and so far it has resulted uniformly in failure…Although I shall continue to resort to it in otherwise hopeless cases in the future, I have become satisfied that it will be abandoned in the near future and assigned to a place in the long list of obsolete remedies employed at different times in the treatment of malignant tumors beyond the reach of a radical operation."

Hall also quotes comments Coley made to the Surgical Section of the Royal Society of Medicine in London in 1909, as rebuttal to two decades of criticism of his medicine and his character: "No one could see the results I saw and lose faith in the method. To see poor hopeless sufferers in the last stages of inoperable sarcoma show signs of improvement, to watch their tumors steadily disappear, and finally see them restored to life and health, was sufficient to keep up my enthusiasm. That only a few instead of the majority showed such brilliant results did not cause me to abandon the method, but only stimulated me to more earnest search for further improvements in the method."

43. Hoption Cann et al., "Dr William Coley and Tumour Regression."
44. In theory, the toxins could have been revived as a therapy, but because they were now listed as a "new" therapy, they'd be forced to go through a lengthy and expensive process of clinical trials for the FDA, hoping for an approval that might never come, for a formulation that would not be proprietary.
45. Hall, *Commotion*, p. 116.
46. As noted in the monograph "Creating Two Preeminent Institutions: The Legacy of Bessie Dashiell," this was aided, once again, with Rockefeller money. There was philanthropic conflict between Coley's appeals for funding for Memorial and John D. Rockefeller's existing support of the laboratories at the eponymous and physically proximate Rockefeller Institute, which had been built in 1901 after young John Jr. urged his father to create something like the impressive European laboratories of Pasteur and Koch. Rockefeller's scientific advisors didn't see much progress coming from Coley's work with "human material." His secretary of the Rockefeller Institute, Jerome D. Green, agreed that their support for Memorial should be discontinued: Instead of funding Memorial directly, John Jr. started cutting checks directly to Coley.
47. From the letters of Helen Coley Nauts, Coley archive of the Cancer Research Institute, as researched by CRI staff science writer Matt Tontonez.
48. So what might have happened differently? What if Memorial Sloan Kettering, under the guidance of Cornelius Rhoads, had not dismissed Coley's appeals and his daughter's letters, and put the resources of "the largest cancer hospital in the world" into a clinical evaluation of Coley's bacterial toxin approach to immunotherapy? It's impossible to say. Medicine in 1950 wouldn't have had

much more of an idea of what to make of Coley's "miracle" cases than Coley himself had in 1890. The immune system was still a mystery. Cancer was still an enemy to be attacked and killed, by weapons developed with a wartime mentality. Immune therapy—using the immune system to fight cancer, as opposed to trying to kill the tumors through toxins or poison—was an idea, but it was no more on Coley's radar than it was on Rhoads's. Everyone was looking for Paul Ehrlich's "magic bullet" that would hit the enemy and avoid the host. No one was looking for a simple stimulator of the body's own natural defense system— a system that was almost wholly unknown in Coley's time and still largely undiscovered in 1950. And the notion of checkpoints like CTLA-4 or PD-1 / PD-L1 were then as unimaginable as they were invisible.

Three: Glimmers in the Darkness

1. In terms of Western medicine, the idea that the immune system could be manipulated to fight cancer dates back to the middle of the nineteenth century, when a German pathologist named Rudolf Virchow described his view through the microscope: a slice of tumor, infiltrated with human immune cells. That was cancer (the tumor), under attack by the immune system.

2. The names for all these biological things tend to get made up along the way, and sometimes before anyone really understands exactly what they're naming, which can make them seem unnecessarily complicated later. For example: the stuff that isn't blood cells in the bloodstream is called *lymph.* That amounts to the white blood cells and fluid. The B and T cells within that fluid became lymph cells, or, with the Greek root for "cell," lymphocytes. Now that refers to both B and T cells, the cells of adaptive immunity.

3. Vaccines have been used ever since Edward Jenner first demonstrated that the immune system could learn and remember. Vaccines safely introduced the body to the unique proteins of a disease it had never encountered before, resulting in immunity. It took several generations to understand how that happened, but even in the eighteenth century it was clear that something in the blood, a few weeks after inoculation, remembered, recognized, and attacked the disease that had been introduced. These discrete chemical bodies worked *anti* the foreign proteins, so they were called antibodies.

4. Biologists knew that these B cells hadn't just been born in the bloodstream. They came from somewhere, maturing somewhere else in the body before migrating to the bloodstream. That place was located in birds before it was found in people. In birds, which have hollow bones, these white blood cells

mature in a sack-like organ wonderfully named the "bursa of Fabricius." B cells are Bursa cells.

5. Three billion might seem like a lot, until you realize that to be fully prepared for *everything*, you need to have 100 million different flavors of antibody, made by 100 million different variations of B cells. That means that when a random B cell does happen to find its antigen match in some bacterial invasion or viral swarm, there are only about fifty other B cells with that antibody to join the fight.

6. David Masopust, Vaiva Vezys, E. John Wherry, and Rafi Ahmed, "A Brief History of CD8 T Cells," *European Journal of Immunology*, 2007, 37:S103–110.

7. J. F. Miller, "Discovering the Origins of Immunological Competence," *Annual Review of Immunology*, 1999, 17:1–17.

8. There would later be found a third type of T cell, which in this metaphor would be something like a referee, regulating immune response and blowing the whistle to stop play and make sure nothing gets out of hand, since "getting out of hand" in the T cell world is dangerous. These regulatory T cells are called T regs.

9. In this ecosystem, cytokines communicate between different immune cells; see the earlier discussion of macrophages. Just to confuse the terminology further, for a time cytokines were called, as a class, "interleukins," either type 1 or 2. They're not all referred to in this way anymore, but we still have all these names kicking around, and so some of the cytokines are still named interleukins and given numbers. But they're all still cytokines.

10. Burnet Macfarlane, "Cancer—A Biological Approach," *British Medical Journal*, 1957, 1:841.

11. L. Thomas, "On Immunosurveillance in Human Cancer," *Yale Journal of Biology and Medicine*, 1982, 55:329–333.

12. Those may include the odd patient in literature going back to Pharaoh Djoser's physician Imhotep. The Ebers Papyrus, ascribed to the physician Imhotep in 2500 BC, advises for treatment of tumors "a poultice, followed by incision," a course of treatment that induced infection. There is some speculation that this treatment may have occasioned an occasional immune response like that witnessed by William Coley. *The Papyrus Ebers: The Greatest Egyptian Medical Document*, trans. B. Ebbell (London: Oxford University Press, 1937).

13. In Europe, records from the thirteenth century describe something similar in the life of a wandering monk named Peregrine Laziosi. He traveled rough, proselytizing and saving sinners as he went. His legs were often sore—the lot of a wandering monk—but at some point he took note that his lower leg was

swollen, and the swelling continued to increase. Soon, a mass began emerging from his tibia. Physicians were consulted and determined it to be a malignant cancer, for which the only treatment was amputation of the leg.

Like many patients, Laziosi heard the physicians' advice and failed to follow through. He continued wandering, the sarcoma continued growing. Eventually the cancerous mass burst through the skin, and the wound festered with infection. "Such a horrible stench was given off that it could be endured by no one sitting by him" (Jackson R. Saint Peregrine, "OSM—the patron saint of cancer patients," *CMAJ*, 1974, 111). But in time his fevers broke and his tumors, remarkably, seemed also to be melting away. Several centuries later the Vatican canonized Laziosi as "St. Peregrin," the patron saint of cancer patients. Where the pope saw a miracle, others saw a potential therapy.

14. Not his real name.

15. According to D'Angelo's file, that was in 1957. He returned to the VA hospital five months later like Marley's ghost, telling his freaked physicians he felt fine. Instead of dying, D'Angelo had thrived. He'd gained twenty pounds. He was working. He had a story but no explanation. It was amazing, what some might call a miracle. But his doctors were sure he was still going to die.

Sometimes cancer does that, growing without destroying the major organs, acting more like a parasite than a disease. It could remain that way for years before it spilled over and became deadly. His physicians assumed it was only a matter of time before reality reared its head.

A year had passed, then another. But when D'Angelo returned three years later with a new lump behind his ear, it was assumed that this was it, the other shoe dropping, statistics borne out. The lump was surely the same metastatic cancer. It must have filled his body and was now pushing the bounds of that body, visible even to the naked eye. This time, his physicians didn't bother opening him up or cutting it out. Once again, D'Angelo was sent home to die. And once again, he didn't.

16. Steven A. Rosenberg with John Barry, *The Transformed Cell* (New York: Putnam, 1992).

17. As a surgeon he was aware of a case where cancer had seemingly spontaneously arisen in a patient with a suppressed immune system, after he received a kidney from a donor who had gone years without showing evidence of the disease. That recipient had beaten cancer, once his immune system was unsuppressed. But this was different.

18. Our bodies were in a "constant struggle for survival and…under constant attack from foreign invaders such as viruses and bacteria," and usually the cells

of the immune system recognize them as foreign and eliminate them. Rosenberg, *Transformed Cell*, p. 18.

19. The extraordinary move from resident to chief of surgery rubbed some of the staff the wrong way, and some referred to him sarcastically as the "boy wonder" or even "Stevie Wonder"—which Rosenberg didn't totally appreciate because he was thirty-four years old and had a family, including two children.

20. Including Dr. Donald Morton, who had been at the National Cancer Institute just prior to Rosenberg's arrival.

21. He was aware of Coley and his toxins, and while he didn't have much intellectual interest in his approach, other contemporary immune researchers whom Rosenberg respected certainly did, including Dr. Lloyd Old, who had identified a substance made by immune cells that he believed had been part of the toxin's mechanism of action (a cytokine he called "tumor necrosis factor," or TNF).

22. Bacillus Calmette-Guérin, or BCG, is a tuberculosis vaccine approved for use as an immunotherapy in bladder cancer.

23. This was Rosenberg's statement, and not as subjective as it might seem here. He pointed out that part of the scientist's job is to sort signal from noise in the scientific literature (Rosenberg, *Transformed Cell*).

24. This was based on work by the English scientist M. O. Symes and on discussions with his friend David Sachs.

25. Rosenberg identified her as Linda Karpaulis.

26. Francis W. Ruscetti, Doris A. Morgan, and Robert C. Gallo, "Selective In Vitro Growth of T Lymphocytes from Normal Human Bone Marrows," *Science*, 1976, 193:1007–1008.

27. Laboratory of Tumor Cell Biology, National Cancer Institute, National Institutes of Health, Bethesda, Maryland.

28. Rosenberg would write, "Ten months after Gallo's paper appeared Kendall Smith, who was at Dartmouth and on his way to becoming the word's expert on IL-2, and his post-doctoral fellow Steve Gillis published an article in *Nature*... about using IL-2 to grow mouse T cells."

29. Several other researchers had also found or described what would be realized to be IFNs previously, but this group is justly credited with the publication: A. Isaacs and J. Lindenmann, "Virus Interference. I. The Interferon," *Proceedings of the Royal Society of London. Series B, Biological Sciences*, 1957, 147: 258–267.

30. The chemical messengers are cousins of hormones, the chemical messengers that quickly and powerfully communicate across the blood-brain barrier and

unleash a menu of cellular changes, depending on the cytokine in question. In the '60s and '70s the discovery of even more of these messengers of immune action and inflammation resulted in a sudden surge of new chemical names and an alphabet soup of acronyms, so many that, as a group, they came to be referred to derisively as "leuko-drek," a blah-blah of immune scientific jargon. Adding to the confusion was the decision by some young immunologists at a conference to take it upon themselves to simplify the nomenclature by referring to all immune hormones as "interleukens" of either type 1 or 2 (depending on the major histocompatibility complex, or MHC: a region of chromosomes that includes a complex, or specific and unique, arrangement of genes that are involved in antigen presentation). That didn't stick entirely (though it has somewhat, hence the confusion). As a class they are now referred to broadly as cytokines.

31. Including into cellular communication, and how signals are translated from the outside of a cell to in, from receptor to nucleus, as shown by the work of James Darnell, Ian Kerr, and George Stark, and others. Ten years after its "penicillin" moment, it would be discovered that interferon alpha and beta were important aspects of immune signaling and stimulation, including the stimulation of T cells. The "failure" of interferon is a misperception that, like many false stories, is far more memorable and harder to dislodge from the truth. And the truth is that IF is now approved as a therapy for use against several human diseases, including hairy cell leukemia, malignant melanoma, hepatitis, genital warts, and others, and it continues to intrigue researchers.

32. For natural IL-2, near twice that for the recombinant form.

33. As an aside: the classic nineteenth-century experiments that resulted in the theory that inherited traits are carried on discrete genes owes much to the fact that Gregor Mendel was not allowed to breed rodents in his monastery, and so he performed his famous experiments using peas. By chance, the genes coding for pea color and the smooth or round surface characteristics happened to be on separate chromosomes, thus enabling the observations leading to Mendel's hypothesis.

34. In addition to a distinguished career in cancer, Dr. DeVita's biography mentions that his son Ted was diagnosed with aplastic anemia and served as the inspiration for the character played by John Travolta in the 1976 made-for-TV movie *The Boy in the Plastic Bubble*.

35. A follow-up to Rosenberg's *NEJM* paper, in the *Journal of the American Medical Association*, provided a clearer picture of those side effects. Of the ten patients

in the follow-up cohort, eight ended up in the intensive care unit. Those side effects included "leaky" blood vessels resulting in extreme fluid retention and grotesque swelling in a very short amount of time, dangerously high fever, racking chills, platelet counts that barely registered, and other various issues that required cardiac catheters, transfusions, antibiotics, and dozens of other secondary medications for this "natural" approach that used Mother Nature's own signaling chemicals and our natural immune defense.

36. Rosenberg, *Transformed Cell*, p. 332.
37. Follow-up NIH-funded trials of Rosenberg's IL-2 results at half a dozen medical institutions around the country failed to replicate his success. There's little doubt that Rosenberg achieved the results his papers stated, and several of his former colleagues I interviewed referred to him as perhaps "the most ethical man" they knew. But the inability of other physicians to replicate Rosenberg's successful results in patients gave many, even oncologists who were immunotherapists, pause when considering IL-2 therapy. A patient's best option, several told me, was to try to get cared for by Dr. Rosenberg personally.
38. Dr. Jedd Wolchok at Memorial Sloan Kettering Cancer Center reduced the approach thusly: "You start by identifying the molecule that's gone wrong, the one that's causing the cancer cell to do the most recognizably 'bad' action—which is to just continue to make more of itself. And then you interfere with that; you short-circuit the pathways, you block it, and make it stop doing that thing." Perhaps the first example of this is the Philadelphia chromosomes, in chronic myeloid leukemia. Another is the BRaff mutation, in melanoma.
39. At the University of Washington, Dr. Philip Greenberg led the conceptual advances of this therapy and was the first to show it could kill cancer in mice. In the following year Rosenberg's lab would work with the T cells that leave the bloodstream and infiltrate tumors (tumor infiltrating lymphocytes); his lab would soon be involved with CARs as well.
40. Rosenberg had used the T cell growth drug IL-2 and showed that by overwhelming stimulation of the T cell army, and overwhelming numbers of T cells, they could push it to kill cancer, sometimes, in some patients.
41. Like the four humors that governed medicine through the Middle Ages, or the nineteenth-century vitalist's belief that living beings contained an ethereal life spark.
42. The modality of virus-related cancers is similar to genetically linked cancers. The virus does not produce the cancer, but it reprograms a cell's DNA to a state where it takes fewer new mutations to line up just right so that a cancer

is the outcome. You can think of it like a slot machine, and a virus or certain genetic conditions as two cherries fixed in place on the dials. The likelihood of that machine hitting three cherries is much greater than in a "normal" machine.

Four: Eureka, Texas

1. On leaving Alice, Allison says, you know, he liked it OK, it was small and he was happy, though—and here he draws a breath, careful, he knows how it can come across—he didn't want to be like every other guy from there. Alice was a dot on a local map an hour west of Corpus Chisti, forty-five minutes if you were hauling it, and it was small-town Texas, a lot of it good—good folks, good farms, a good upbringing and jobs, being close to the air force base. His father had earned his wings as a flight surgeon in the reserve there, and he translated that to becoming the local physician, but his people had been from Waco, where they'd owned a shoe store. He'd made a leap to Alice. Jim wanted the next leap. He wasn't going to be happy there. He'd grown up in a small town, he loved it, loved the place and the people as you can only love what you're from. And knew them, too, as only a local boy can know a thing. And, he says, he didn't want to just be that guy, like every other guy.

   There was nothing wrong with it. He did like the football, just not playing it; he liked the small-town vibe, just not being stuck in it. He was a reader, a tinkerer. He had a curiosity and an idea about himself, too, not necessarily precocious, more like potential. The family garage became a lab, the woods a place to test homemade black-powder bombs, the ponds a source of dissectible amphibians. If you talk to most research scientists, have a couple beers with them, you realize they all were a little on their own, and many made homemade explosives; it's normal, young-scientist stuff. And if that made him a weirdo in meat-and-potatoes Texas, so be it. There were three boys in the house anyway, and the two older ones more than made up for it. His dad supported him when he lobbied to take an advanced high school biology correspondence course through the University of Texas, rather than suffer through a senior year of biology taught without evolution. And then the following year he turned sixteen and graduated, and he was gone for good. Austin was the center of the local freak scene, and it had a great university, too. But on both accounts, eventually he found out Berkeley was greater and freakier than almost anywhere.

2. These were the peak years, when it was still wide open, a little university town just beginning its metamorphosis into the freak capital of a cowboy state.

3. Of course, Jim Allison wasn't alone in that. These were Austin's boom days. The town had been a tiny university town until the baby boomers graduated high school. Smaller than big cities like San Francisco, which received and fostered the flower power of the '60s, landlocked and deep-Texas Austin became the repository of the local version, still Texas enough to two-step, hippie enough to do it stoned, and close enough to the university to ensure a steady flow of bright young things with an eye on the future. For many, that future no longer required heading to bigger cites on the coasts, or overland to Dallas or Houston. It could be found right there in Austin, where Texas Instruments, Motorola, and IBM has recently relocated their manufacturing.

   Of course, it didn't hurt when the voting age—and the drinking age—was lowered to eighteen and the closing hours were extended to 2 a.m. Politically empowered, legally drunk teenagers could thus romp past midnight in the freak capital of a cowboy state, and a music scene quickly sprang up to provide the soundtrack. If you sold beer and had a surface flat enough to put a bar stool on, you were a music club.

4. "I was trained in biochemistry, so I was working with asparaginase, which is this enzyme that can deplete your plasma of asparagine, which many leukemias need to grow. They can't make their own, enough. It's still used to induce remissions in childhood leukemia, but it doesn't cure anybody. In mice it cured the leukemia. I was trying to work on making it work better. I began to read about this immunology stuff. I took a course and I got really excited and interested about it. Just for the hell of it one day, I cured mice with this enzyme of this leukemia that they had."

5. When injected into the mouse, the enzyme broke down the fuel. As a result, the leukemia was starved and the mouse cancer cells died. Then those cells, like all cells that die in the body, were cleaned out by the roving, garbage-eating cells of the innate immune system, the macrophages. Allison wanted to know how they worked, too—how all of it worked, really.

6. His voice has the distinctive music of the back country, a little extra beat put into the last word, seemingly no rush to move to the next one, except when that beat is over he moves on quickly, where he was going all along.

   "I was lucky enough to be at one of the very few universities not connected with a medical school that even had anything on immunology," Allison says. He was training to be a biochemist, but the work had sparked an interest in another aspect of biology—the immune system—and one of his graduate professors, Jim Mandy, offered a course. Allison jumped at the chance, "and I was just fascinated by it." His professor lectured on the newly discovered T cells. "He taught the discovery," Allison says. "He gave the lecture. But after hours you go see him in his office, he'd tell you he didn't really believe it was true. He was an antibody guy." What bothered Allison's professor—and a great many others in immunology—was that T cells seemed too different from B cells to be part of the same system.

B cells didn't kill disease directly; they made antibodies, and the antibodies marked the disease to be killed by the innate immune system. That had been immunology for years, and the direction of research was to continue to clarify that scenario. "But these T cells came along and people were saying, 'Well, these work differently, by killing infected cells directly,'" Allison says. Adding T cells to the B picture seemed too complicated. Evolution tends to be a conservative force, using the same biological processes again and again, repurposing and building on the biology it already has rather than starting from scratch. If the immune system was complicated, those complications were most likely to have grown out of common roots and to utilize similar mechanisms. It was almost beyond imagination that nature would have evolved two totally different types of systems with overlapping jobs in the same organism. "He taught it anyway, but then I'd go and talk to him in his office and say, 'Dr. Mandy, why don't you believe T cells kill infected cells?' He'd say, 'Well, I just don't know, it just seems too weird, you know?'" It was as if each of our kidneys removed toxins from the blood in completely separate ways, with no relation between the two. Allison thought it was weird too, good weird. He wanted to "check it out," learn more about it. "It was a fantastic time in science," Allison says. "Immunology had just always been this poorly understood field—I mean, everybody knew we had an immune system, because there were vaccines. But nobody knew much about the details of anything." He'd already hit the ceiling in the only immunology class available in Austin.

7. The macrophages and dendrites (biologists just say "antigen-presenting cells," or APC's) act like a living billboard showing the latest winning lottery numbers, in the form of unique samples of disease antigens. Every one of the billions of adaptive immune cells was born with a different lottery ticket. Sooner or later the number hits: the B or T cell happens to randomly match exactly with the antigens being reported, and they start multiplying into a clone army of themselves, all with the same winning ticket against the disease being shown, and bingo, adaptive immune response is initiated.

8. "It was kind of disappointing," Allison says. He had wanted to go somewhere "first-rate" for his immunology education. "But I was just doing biochemistry

again, purifying proteins and sequencing them and all this stuff. The older guys would just have you doing this grunt work, everything else they called model making, as in, 'don't make models, don't think, just do the work!' So I said, if this is science, you can have it, I'm going back to Austin! But at the time I was in San Diego. I was married and I played with a country and western band a couple nights a week. I had a pretty good time.

"Remember Spanky and Our Gang that did that song, 'Like to Get to Know You' and all that stuff? I played with a band that opened for them one night and actually sat in with Spanky McFarlane." A longhair PhD harmonica player was cool, an easy fit in the music scene.

"Well, I played at this place called the Stingray. We had a band that played. It was called Clay Blaker, the Texas Honky-Tonk Band. I had a day job. They didn't. Everybody else in the band would...I'd just hang out and play, you know, a couple of sets, maybe, or half a set every now and then, or something. I got to know them really well. Through them I got to know other people. I was pretty popular as a harmonica player for up-and-coming singers that would want to play at open mic night and stuff like that. Our band got pretty famous in what was called the North County. This was up in Encinitas, California. I got to see a side of life that I've seen a little bit when I was growing up in Alice, but you know. I mean, pretty rough and tumble."

At the time, Jim was married and working seven days a week, and, as always, playing hard. "We played every Tuesday night and a lot of Friday nights and sometimes nights in between." It was rowdy sometimes, playing harp in the country-western dives in the sticks. "People don't realize, that part of California is pretty redneck," Jim says.

"There were fights pretty regularly, usually it'd start because one cowboy that's doing a two-step would swing too widely and bump into a guy and the guy would say, 'Don't do that again.' Then, that's just the way the guy danced, you know? So it happened again. Add beer and a crowd and pretty soon...

"Actually, it was this guy named Luther, he was one of the guys that came to see us all the time. We really liked him. That's who he was. He was just this big, gangly guy that danced big. There were some other guys from this other club that heard us play at another club, so they decided to come to our club to see us. It was almost like a gang or something.

"After three or four times one night, the guy slams into Luther. You know, Luther was everybody's friend. I was up on the stage and there was a guy there, it was so crazy, it was a guy that had just gotten...He was pretty rowdy, but he had spent some time in jail for stealing horses. Anyway, he was there and he'd broken his arm somehow and it was in a cast. He came running up to this guy that had hit Luther and, you know, hit him with that cast and the guy just went parallel to the floor. I was playing, you know? The guy goes parallel to the floor and I jumped out of the way, you know? The guy jumps up on the stage and now it's like something out of a Western.

"I went, 'Whoa.' That guy down there who I knew pretty well, but he was just going, 'Oh, shit. Oh, shit.' There was stuff like that. It was a lot of fun."

One night, he tagged along to a musician's party—crashed it, really, since it ended up being a release party for Willie Nelson's new album, *Red Headed Stranger*. That led to taking Willie Nelson and part of his band to an open mic night at a honky-tonk, then back to their hotel in his faded red VW microbus.

Many years later Allison would end up standing in for Willie's harp player. Allison is a founding member of an all-immunologist band called the Checkpoints. They really are pretty good.

9. "If I saw something interesting, you know, I'd chase down a couple of the things they'd cited and Xerox them and take them home.

"At that time, I was living in Austin. My wife worked in Austin and I was living there and commuting forty-five miles a day to Smithville. Ultimately, we bought a house in a failed development with eighteen acres of land. The lab was in a state park. It was in the woods in a clearing. I bought in the woods, about a mile and a half, because I had a motorcycle, or I'd walk sometimes through the woods. Then, I'd go back to Austin on the weekends just to party."

He didn't have time to gig out then, he was too busy, but he could still catch Willie Nelson or Jerry Jeff Walker at the Armadillo Worldwide or the Soap Creek Saloon.

10. Soon after Allison joined the team, the president of MD Anderson left. "The new guy came in and didn't really know who we were."

11. By then it had been shown that there's major histocompatibility complex (MHC) restriction. T cells don't recognize just an antigen; they recognize it in the context of these MHC molecules. The MHC molecules are a distinct arrangement of proteins that can be thought of something like blood type, in that we're all born with one or another of a few flavors, genetically determined. Not every person shares the same MHC, but all the cells in a person's body share the same group. The MHC complex is sort of like a tribal mark or signature on every cell's surface, and it serves as a basic but effective factor that allows the immune system to be a better shepherd, to keep track of what is us, and to recognize what is a foreign invader. (It's also what needs to "match" so that tissue or bone marrow transplants won't be rejected.) Allison had been

studying and working with experiments involving MHC molecules in his lab at Smithfield, and he'd been obsessively following the latest developments in the immunology journals. He knew the MHC was an important factor in how this mysterious T cell receptor worked—a factor other researchers seemed to be ignoring.

Jim had a different kind of molecule in mind as the T cell receptor, and he'd thought of a different sort of experiment to find it.

12. Finding it was like looking for cilantro in a field of parsley—in the dark.

13. People were looking for immunoglobulin chains that were made by T cells.

14. Including immature thymocytes.

15. The experiment staked a solid claim, but it wasn't absolute proof that Allison had found the holy grail. His experiment was just that, an experiment, providing results, and Allison didn't have the sort of pedigree that provided him benefit of the doubt. "Nobody believed it because I was this guy in Smithfield, Texas, you know?" Allison makes clear that whatever his experiment did, it didn't "prove anything. Science rarely, if ever, proves something, but good science can present good data, and good data can strongly suggest."

16. Academic papers follow a standard, dry format that lets the data speak for itself. It's in the "discussion" section at the end where authors may speak more personally, if imprecisely, about further implications that might be suggested by the data. Allison's paper, titled "Clonogypic Antigen of T-Cells," followed this format. The text was dry and factual and made no claims, carefully explaining what he'd done without mentioning "T cell receptor" at all. He made up for it in the discussion.

17. Like Davis, Tonegawa had been working on unraveling the genetics of immunology since the mid-1970s, and Tonegawa had been first to identifiy the gene in B cells that allows them to make millions of varieties of antibodies to meet a wide diversity of pathogens—a goal that Davis had also been working toward.

18. Chien et al., "A Third Type of Murine T-cell Receptor Gene," *Nature*, 1984, 312:31–35; Saito et al., "A Third Rearranged and Expressed Gene in a Clone of Cytotoxic T Lymphocytes," *Nature*, 1984, 312:36–40.

19. Davis would later recount for a reporter for *Stanford Medicine* magazine that the editor of *Nature* had called him to recount how unhappy Tonegawa had been at this "divine justice," but he said his MIT competitor had been magnanimous in defeat.

20. This was the sort of thing Allison dreamed of as a kid. Was it for Mom? Jim says no, but that's a maybe. Maybe everything we do is for Mom, one way or

another. If anyone knows, it's certainly not the doer. But it's true that he had that experience and it stuck with him, just eight or ten as he was, and so, maybe he wasn't eight or ten—man, it gets mixed up, you start talking about your mom and dad.

What he knows was, she died. He was there, he didn't know what the disease was or how you fight it, then later, he learned what the disease was, and that nobody had much useful to say about fighting it, and he thought: Fuck this. I'm going to do something.

21. Justifying himself in front of fifty of the top scientists in the world was no picnic, and even the memory of those visits, which were timed to the second by stopwatch, still knots his stomach. "It was pretty bad," he says. "Sometimes the night before I'd just be in the bathroom throwing up." But the trade-off was, Allison finally had all the resources he could ever need to get down to work.

22. "It wasn't my idea," Allison clarifies. "It came from a guy named Ron Schwartz with NIH and a postdoc in the lab, Mark Jenkins. They showed that just the engagement of the antigen receptor itself was not enough to turn on a T cell. And it showed increased deselectivity." See Mark K. Jenkins and Ronald H. Schwartz, "Antigen Presentation by Chemically Modified Splenocytes Induces Antigen-Specific T Cell Unresponsiveness In Vitro and In Vivo," *Journal of Experimental Medicine*, 1987, 165:302–319.

23. Allison had seen it himself, and experiments at the NIH had proven it. Allison had spent years on the biggest, most complex jigsaw puzzle in biology, and this new revelation was asking him, and everyone else in biology, to rescramble the pieces. Which, Allison thought, just made the whole thing "more interesting."

24. "Only certain cells could do that. Later on it turned out they were dendritic cells that Ralph Steinman [whose lab included a young Ira Mellman] got the Nobel Prize for a few years ago. So we did a lot of work showing where they came from, never could show what they did though."

25. "Anyway, then I got into the idea of combination-stimulation, and a second signal came along. So we had the lab dive into that, so we came up with the whole idea that CD28, I mean there is a molecule that a lot of other people worked on. Yeah, so a bunch of people, Jeff Ledbetter, and Peter Lindsley, Craig Thompson, and some other people have been studying. They made an atom out of this thing called CD28, and it would partially activate T cells. There was a lot of literature that it would do things in humans but this issue of a second obligatory signal, their work really didn't answer it. Partially because you really can't, it's hard to do with human cells, because humans don't have a lot of, one reason,

humans don't have a lot of naïve T cells. Because, we've had so many infections overall, so in your blood most of the cells are there looking for something to do. We've seen it before where mice, we keep them clean."

26. "A guy named Jeff Ledbetter had really been studying [it] for a long time, along with Craig Thompson and Carl June and Peter Lansing and others," Allison explains. Allison had a couple reasons to think it might be the costimulating signal.

27. Allison did the experiments, and they worked. "So, it's officially necessary to give that second signal," he says. And that seemed to be that. He published the paper. "I was really happy," he says. "I'd been working on it for like three years" he says, "just thinking about it—everything moves slow." But Allison was a researcher. And thinking about CD28 had led him to thinking about the unique problem of cancer. It didn't get attacked by T cells. Most scientists assumed it was because this was a self cell, too similar to normal healthy body cells to be recognized by the immune system. But Allison now had different thoughts. And as it happens, he was having them at the very moment when he was doing basic research in a well-funded cancer lab. "It occurred to me that, since tumor cells didn't have those [CD28] molecules, maybe they're invisible for the immune system, even though they have tons of antigens. The immune system can't see them, because they can't give that second signal."

28. "So that was a science paper, but all the time that was going on, when we cloned mouse CD28, we didn't first identify the molecule. Other people did, even cloned it, but the human one. We cloned the mouse T and studied it and showed that CD28 was this costimulatory molecule."

29. Signaling proteins have an inside-the-cell and an outside-the-cell aspect; they stick out of the surface through the cell membrane like a carrot sticks up from the ground. The outside part interacts with the outside world and receives the signal. That signal travels through the protein to the inside of the membrane and the aspect of the signaling molecule that is inside the cell, which is where the action happens; then it initiates gene expression, a sort of "reaction" to the signal. What Allison and Krummel found was a molecule in the gene bank that had an exterior component—the carrot green—that was "like 85 percent identical" to the exterior signaling portion of CD28. That family resemblance might have been coincidental, but Allison felt that the better bet was that it meant the two signaling proteins were in fact closely related evolutionarily—and did similar things. "To me everything comes back to evolution sooner or later," Allison says.

30. "This guy Chip Holstein had cloned it," Allison says, which allowed researchers to study it further. "Didn't know what it did, just knew it wasn't in naïve T cells that got turned on."

31. "CTLA-4 comes from this guy, Pierre Goldstein from France, who was again doing subtractive hybridization to find things that were expressed only in T cells. He took T cells and subtracted RNA that was in B cells as well and looked at what was left. The fourth thing he got was CTLA-4, or cytotoxic T-lymphocyte-associated antigen protein #4. That's where it came from, and it's a complete misnomer because it turns out, it's in all T cells, not just CTLs (killer T cells). It's also in helper cells. It's in every T cell after they get activated. But I like the alliteration of that, CTLA-4." It's also called CD152.

32. Linsley et al., "Coexpression and Functional Cooperation of CTLA-4 and CD28 on Activated T Lymphocytes," *Journal of Experimental Medicine*, 1992, 176:1595–1604.

33. Krummel designed a model to press both pedals and tried it in animals, then dialed the mix of gas and brake, CD-28 and CTLA-4, like a new driver. Demonstrating that you could drive T cell response up and down in animal models as he'd predicted on his spreadsheet, made with an early version of Excel, really drove home that they were looking at what they'd suspected. "Jim was really a hands-on PI at that time," Krummel remembers. "I think he taught me how to inject my first mouse." Krummel says Allison's attitude for his chosen postdocs was essentially *Trust your instincts. Try things.* He also paraphrases this as *Fuck it, try it.* Krummel tries to instill that spirit in his students as a professor of pathology with a laboratory at the University of California, San Francisco. "I didn't even know anything and was allowed to throw antibodies into mice," Krummel says, by way of a tribute to the culture of Berkeley at the time and Allison's lab in particular; certainty in science is often anathema to pure exploration.

34. The belief at the time was that T cells were calling the shots in terms of this immune response. It is now believed that macrophage cells from the innate immune system—those large, hungry "garbageman" cells that gobble up the detritus of the body help regulate immune response by means of cytokines. It's also now understood that T regs, which had not been discovered at the time of this CTLA-4 work, are the cells primarily expressive of CTLA-4 and so play an important part in down-regulation of activated T cells.

35. "Jeff Bluestone, who was at the University of Chicago just about the same time, independently did the same thing," Allison says. Bluestone was an immunologist (he's now the CEO of the Parker Institute for Cancer Immunotherapy) and his lab was trying to utilize this newly discovered immune brake to

prevent rejection in organ transplants and autoimmune-related diseases—issues that were widely accepted as being within the immune system's wheelhouse.

The majority of cancer experts and immunologists believed that cancer had nothing to do with the immune system. Allison, meanwhile, was a biochemist who had wandered into immunology, and, in a pattern that repeats itself throughout the history of advances in cancer immunotherapy, he wasn't aware enough of the battles between believers and nonbelievers in cancer immunotherapy to realize that he'd casually stepped across the battle line. The next step in his experimentation was more controversial.

36. And so, even as he had his eyes fixed on the road ahead, doing pure science on T cells, that thought was always traveling with him in the passenger seat. He describes it sometimes in terms of a Jerry Jeff Walker song about a cowboy cruising down the highway with one eye on the road and one eye on the gal next to him. Even as he was driving, he's always angling on when he can pull over.

37. Allison knew cancer. He knew it as a kid, though they didn't call it that. Cancer wasn't a word you said then, it was a dirty thing, a curse, the C word. You didn't say it, but Jim could see it. It was in his mother's eyes, the way her dress hung as she set the table, hiding her exhaustion in silence and forced smiles. That was Texas and she was cow people, tall boots and tall cactus and family stories of the Chisholm Trail. Cow and horse people, true Texans, and not given to complaining, not even with the disease progressing unchecked or her pale skin flushed from the radiation burns, the only way science had to arrest it. That was how it was, three summers and getting worse but not getting talked about. Jim remembered one summer day when one of the adults came to find him and told him he needed to go home, now. Five decades later he can still feel her hand go limp and the light go out; that stark, awful moment when it changed still brings water to his eyes. So was this work, this needling thought in the back of his mind, for his mother? Maybe everything we do is for Mom, one way or another.

"Well, I didn't know why she had it, I just knew she was sick. Nobody talked about cancer, nobody said cancer, nobody in my family, I didn't know what was wrong with her. I didn't know what cancer was. I just knew my mom was sick. One day, I was headed to the swimming pool with some friends and somebody came running out of the house and said, 'No, you can't go. You have to come back and be with your mom.'

"And I still didn't know what it was. I mean, I was holding her hand when she died. And I didn't know what it was, I just knew she was dead. I had to put it together later because I was just too young to know. It pissed me off, though."

Alice wasn't a big town, and the Allisons lived on the edge of it. They were already on the edge of outside. Jim's mother's death pushed him over that edge. He spent a lot of time just walking, not knowing here he was going, kicking the dirt, keeping his feet moving so his head couldn't think too hard about what had happened. That was how he found the settlement.

He'd stumbled into it by accident, a decaying little ghost town in the woods. He was picking through the woods, trying not to think but thinking anyway, and suddenly he realized, this was a place. Somehow, it still was. Here, among the wet leaf floors and creeping walls of moss, watched only by the damp eyeholes of collapsing root cellars and the ghosts of failed generations of dirt farmers, Jim Allison dreamed of something bigger. Genetics, environment— the causes of cancer mattered, from a scientific perspective; everything mattered. From a personal perspective it didn't matter worth a damn. Cancer was just true, whether you understood it or not. You're here and you're gone, like these families, like this town, falling back into dust. It was his mother gone to that thing. Later it took his older brother too, prostate cancer this time, and shortly after, when Jim got the diagnosis of the same thing, he just said, "Fuck it, cut it out."

38. In fact, it would make it through, in 2011. Once approved by the FDA it was marketed under the trade name Yervoy; it would cost patients $120,000 for a full course.

39. Allison and Krummel both appear on the provisional patent application. Allison's postdoctoral fellow Dana Leach would also be added.

40. Now known as the Geisel School of Medicine.

41. These mice had been genetically engineered; their immunoglobulin (antibody protein type) genes had been replaced with human immunoglobulin genes. As a result their immune response against a foreign protein (in this case a human CTLA-4 receptor) produced antibodies made of proteins that wouldn't look foreign to a human, and those antibodies could go into humans without triggering an immune response against them.

Five: The Three *E*'s

1. It's important to emphasize that Allison consistently makes clear the contributions, many of them essential, of those in his lab. He names Dr. Matthew Krummel in particular, and is even more clear on the fact that Dr. Jeff Bluestone—who at the time had moved from the University of Chicago to the University of California, San Francisco—had simultaneously discovered that

CTLA-4 was a down-regulating signal, an immune brake rather than a gas pedal. Bluestone has been consistently and publicly recognized for this work, but because he applied it to further research on the down-regulation of immune response, rather than blocking that down-regulation with an antibody and testing it against cancer, his name is not as widely associated with the breakthrough in cancer specifically. Bluestone is currently the president and CEO of the Parker Institute for Cancer Immunotherapy, a job that has him at the center of funding and coordination of the efforts of thousands of scientists and researchers around the world.

2. Dr. Old's 2011 obituary in the *New York Times* credited him as being essential to an approach to cancer "also known as biotherapy."

3. He was an heir to Coley, personally championed by Coley's daughter, and, somehow, also friendly to the cancer gods at Memorial Sloan Kettering Cancer Center, where he maintained a lab and office and held the William E. Snee Chair of Cancer Immunology.

4. As frequently you'll hear him referred to as "the father of tumor immunology." In addition to articulating the concept that cancer has unique molecular "ID tags" (antigens) that should make it a unique target for the right kind of immune response, Old was responsible for such major immunological advances as the discovery of a legitimate bacterial treatment heir to Coley's Toxins, in the form of bacillus Calmette-Guérin, or BCG; it was one of the first immunotherapies approved by the FDA and is still effective against some forms of bladder cancer. He believed in an immune-based interaction between cancer and the immune system, and kept the field alive during its darkest days. He was a remarkably broadly educated man, a concert-level violinist, an accomplished immunologist, and the heir apparent to the torch borne by William Coley. He was also the founding scientific and medical director of the Cancer Research Institute, founded by Dr. Coley's daughter, and he held this position for more than forty years. By reputation and as relayed through dozens of interviews, Dr. Old was part Osler, part Huxley, and all mentor. Unfortunately Old died of prostate cancer in 2011, at the age of seventy-eight. His death came just before the approval of the first checkpoint inhibitor.

5. Even in the darkest hours, when researchers had mustered little solid scientific data to support the theory, Old remained an unapologetic believer in the interaction of cancer and the immune system, and he worked to popularize and explain those ideas through articles in both scholarly peer-reviewed scientific journals and the popular press; his 1977 *Scientific American* article, rakishly titled "Cancer Immunotherapy," lays out the basic concepts for a lay audience.

6. Robert D. Schreiber is the Alumni Endowed Professor of Pathology and Immunology at Washington University School of Medicine. Bob was also kind enough to remember me. He'd have had every right not to. I'd been sitting at the bar of the Copley Hotel in Boston during a typically crappy Boston winter day. There was a cancer immunotherapy conference in full, florid bloom in the surrounding conference center, and college football games on a series of distantly hung televisions. There were potted ferns, and there was an open table where I'd stopped in to say hi to Jim Allison. What Jim Allison had been talking about regarded his landmark discovery of the thing you could turn off and on in humans and cure cancer. Then Jim had introduced me to Schreiber and said, with no hesitation, "I found it, but Bob proved it." I wrote that down and wrote down his name and about a year later finally got around to figuring out what the heck Jim Allison was talking about.

7. They weren't mice, but they were close cousins, and you could use them in mice, so their immune systems didn't reject the antibodies as foreign. "They were non-immunogenic. So you could do in vivo experiments even before the days that you could easily make knockout mice," Schreiber says. Incidentally, using Armenian hamsters was not the norm for most biologists, since mice are the usual experimental animal, the lab rat of imagination. Schreiber had first read about Armenian hamsters in a journal article. He and his colleague, Kathy Sheehan (Kathleen C. Sheehan, PhD, cohead of the Immunomonitoring Lab and assistant professor of pathology and immunology), tracked some down at a lab in Brandeis University, where a researcher had been using the same inbred population for so long that he'd essentially created a standardized genetic population. The outcome was that they didn't produce antibodies in mice—which turns out to be critically important here, because we now realize that, especially in terms of studying immune response, what works in mice does not always translate to humans, and vice versa. Specifically, much of cancer immunotherapy does not work in mice—which turns out to have been yet another hidden wrench in the gears of scientific progress in that field.

On a related note, penicillin, which has saved many millions of human lives, is fatal to mice. Luckily, penicillin was discovered and quickly tested directly in humans as part of the war effort. If it had been subjected to the usual protocols for FDA approval and gone through mouse models before human trials, that breakthrough might not have been recognized, and many millions of lives would have been lost.

8. The work might sound like developing a key and throwing it toward a lock, and that's not totally wrong, except for the seeming randomness of the process.

Here, you make a key you think fits a lock, and you put the two together in the lab. If the lock fits, that proves something—but what does *fits* mean? As always, the metaphors help, but sometimes they confuse as well. For example, you might think that a "fit" between a key and a lock would turn the lock mechanism and activate the lock. In fact, it's something like the opposite. If the key fits the keyhole, it blocks the keyhole; it prevents the lock from functioning. The metaphor is more like a parking space. If you block it, nothing else can use it. Bob's lab had discovered that their molecules fit the cytokine keyhole.

9. According to Schreiber's recollection.

10. The experiments were based on a guess—or, in scientific terms, designed to attempt to refute if they could, and support if they couldn't, a specific hypothesis. The hypothesis was that interferon gamma made a tumor look even more foreign (immunogenic) to the immune system, and therefore played an important role in amplifying the immune response. Experiments were the only way to test those guesses.

Bob thinks back on it and smiles. "I said, well, I thought—maybe there's a sort of amplification system," he told me. "One that happens between gamma interferon and TNF." Bob thought interferon gamma maybe amplified the signal or the result of TNF. Maybe interferon gamma, somehow, made that tumor easier for the TNF to recognize. "So," Bob said. "Wouldn't it be interesting if in fact what gamma interferon was doing here is actually affecting the tumor, to make the tumor more immunogenic?"

In the chain of dominos, interferon gamma would be in the middle there, a domino that falls and hits two more. Then each of those falls and hits two. You could see it as amplification, or you could see it as a safety mechanism— this was the immune system we were talking about, the double-edged sword that fights measles and manifests as AIDS. The immune system had to be randomly ready to fight anything, including things it had never encountered. It couldn't have a great number of random answers at the ready, of course, but it had to have at least one that could recognize the new random threat, one for each. Then it needed to be able to turn that one soldier ready to fight that random threat into a whole army. But it also needed to make sure that it fought only threats. It was about amplification and modulation; the immune system needs to have both amplification and safety in order to make an attack signal strong enough to effectively communicate the need to join the fight, but conservative enough not to cry wolf and trigger a cannibalizing attack by the immune system on its own body.

11. "It's an inactive form of the IFNγ receptor," Schreiber explains. "That mouse has a lot of trouble mounting cell mediated immunity. So it has a significant deficit, and by that criteria is immunodeficient."

12. As it happens, one of the students in Bob's lab had recently figured out a way to make a mouse that expressed a dud form of interferon gamma—meaning, interferon gamma that plugged in but didn't turn on. They transplanted Old's tumors into the dud interferon gamma mice, and they transplanted them into normal "wild-type" mice. Then they gave both mice tumor necrosis factor. In the normal wild-type mice, the TNF killed the model tumors. In the mice with inactive interferon gamma, it didn't.

13. Reputable hardcore immunologists didn't spend too much time thinking about tumor immunology. Few scientists did. As a result, the few in the field who did get results were looked at suspiciously. It wasn't that they were considered charlatans or wizards, but their results weren't always reproducible in other people's laboratories.

14. Really, nothing is ever fully answered or proven; theories are supported, evidence is presented to suggest conclusions, and data suggests what we might call answers to questions. But to assume that any question is ever completely and definitively answered is to ignore the history of science.

15. Ehrlich was exceptionally prolific and is considered the father of modern immunology, among other fields. As Arthur M. Silverstein points out in his second edition of *A History of Immunology*, Ehrlich had worked in Robert Koch's laboratory in Berlin, and in addition to his medical studies held a lifelong interest in the relationship between the structure of molecules and their biological function. This interest and insight on structural chemistry made him uniquely qualified to postulate the physical stereochemical relationship—and the unique binding affinity—between antigens and antibody. The fuller extension of this line of thinking—his conception of the perfect medicine—is the foundation of the mechanism of immunity, and much of our drug delivery. Ehrlich posited that if one could make a molecule or compound that was attracted only to a pathogen or diseased cell, then that molecule would serve as a guided missile—or, in the language of nineteenth-century technology, a "magic bullet" (*magische Kugel*), guiding any poison payload uniquely to that disease while sparing the host.

    To that end Ehrlich's lab tested hundreds of different compounds against various disease-causing bacteria. Eventually, in variant 606, he discovered one that was safe in humans, but a deadly poison for the spirochete responsible for syphilis. The resulting medicine was called Salvarsan, a transformative medicine for which Ehrlich is best known and a contributing factor in his receipt, along with Élie Metchnikoff, of the 1908 Nobel Prize in Medicine and Physiology.

After Ehrlich's death in 1915, the street in Frankfurt where his famous laboratory made this discovery was renamed in his honor; it was renamed again during the rise of the German National Socialist Party's systematic attempt to erase its Jewish citizenry from national memory.

16. Commercially available laboratory mice are a relatively recent phenomenon, and most come from the grounds of the Jackson Laboratory in Bar Harbor, Maine, on Mount Desert Island. The modern lab mouse has roots in the various strains favored by late-nineteenth- and early-twentieth-century "mouse fanciers" as exotic pets, and are a genetic mix of four distinct and geographically disparate mouse subspecies: *Mus musculus domesticus* (from Western Europe), *Mus musculus castaneus* (from Southeast Asia), *Mus musculus musculus* (from eastern Europe), and *Mus musculus molossinus* (from Japan). According to Jackson Laboratories, many inbred mouse strains originated in the early-twentieth-century colonies of Miss Abbie Lathrop, a mouse fancier and breeder from the dairy land of Granby, Massachusetts.

17. These mice are also called "athymic" or "thymus-lacking."

18. In January 2018, researchers affiliated with the Parker Institute for Cancer Immunotherapy announced the discovery of a molecule called BMP4, which, in mice, helps to promote thymus repair and even regeneration of the organ. The results, published in *Science Immunology*, were developed in the lab of Dr. Marcel van den Brink at Memorial Sloan Kettering Cancer Center, in collaboration with Jarrod Dudakov at the Fred Hutchinson Cancer Research Center. BMP4 will next be explored in humans, with the possibility of drug development for revitalization of the organ and the attendant quality of T cell response in humans. The thymus may be damaged by disease and is diminished as we age and is theorized to perhaps be related to the reason older persons are more susceptible to certain cancers. See Tobias Wertheimer et al., "Production of BMP4 by Endothelial Cells Is Crucial for Endogenous Thymic Regeneration," *Science Immunology*, 2018, 3:aal2736.

19. Timing is everything in such experiments, and it's important not to accidentally cast a scientist doing good science and conducting hard skeptical tests on scientific theories—as scientists are supposed to do—in the light of a villain. Stutman used nude mice, athymic mice. He was right that they lacked a thymus, right that the thymus was where T cells matured, and right, even in 1974, that those T cells were responsible for adaptive immune responses. But what

Stutman didn't realize—nobody did at the time—was that these mice still had other cells from the nonadaptive immune system, called "natural killer cells." They are somewhat the grunts of first-line, basic immune defense in the body, nothing like the trained elite special forces of the T cell army, especially the "serial killer" CD8 killer T cells—but they are present, and can kill modest and obvious invaders. Meaning, he hadn't obviated the possibility that immunosurveillance was in fact still intact in his experimental mice. Perhaps more important, the specific genetic strain of nude mouse that Stutman had used was explosively susceptible to developing tumors from the carcinogen he had used. His mice might have been overwhelmed by tumor development, to which no level of immune surveillance could have kept pace.

20. Osias Stutman, "Delayed Tumour Appearance and Absence of Regression in Nude Mice Infected with Murine Sarcoma Virus," *Nature*, 1975, 253:142–144, doi:10.1038/253142a0.

21. It would later be discovered that nude mice aren't quite as nude as you'd think; they do have small numbers of T cells and "natural killer" cells, whose role in immune surveillance is still unclear. Also, the nude mice strain Stutman had used would later be discovered to be especially susceptible to 3-methylcholanthrene, especially at the large doses Stutman had used, which would cause cancerous mutation in even the strongest mouse immune system.

22. It was possible to build a mouse in which they'd knocked out the gamma interferon receptor. Or they could make a mouse that lacked a signaling protein necessary for gamma interferon to function. They had already made that second mouse in Bob's lab. Or, another way to get there, a knockout mouse. They could use a mouse that had no lymphocytes—no B cells or T cells, and therefore, no adaptive immunity. They had some of those mice available, too, what were called "RAG knockout mice," in which the gene for lymphocyte production had been knocked out genetically.

23. Key to this experiment's quality—the lack of "garbage in"—was ensuring that the carcinogen used was not one that all of the particular breed of mouse being used in the experiment had carcinogenic affinity for, as Stutman's had, and ensuring that the amount of carcinogen delivered was a minimally efficacious dose for tumor development. Stutman had unwittingly overwhelmed his mice with a cancer that no immune system, intact or otherwise, could match.

24. "Another issue was, we were confronted with the argument, 'Hey, I'm a tumor biologist, and I make oncogene-driven tumors, and I never see a role for the immune system in my oncogene-driven tumors,'" Schreiber says. "We only

discovered recently that these oncogene-driven tumors—which are the experimental model of tumors—don't develop any mutations, or at least few. So if they're not particularly immunogenic, it's only because they don't have neoantigens."

25. Schreiber: "You know, you can have deletion of your tumors, elimination. You can have modification of your tumor, so it might be held as sort of an idea, as a…dormancy that we call equilibrium. And it could be altered in such a way, just as you might alter a manuscript, that it would come out as a better tumor."

26. That became a *Nature* paper, too.

27. "We began to look at the tumors that we had passaged in vivo and chart their progression versus regression growth characteristics and using a genomics approach."

28. "One tumor had a very strong mutation in a highly expressed protein. The protein was present before we put it in the in vivo passage [i.e., before it was transplanted into the living animal], but then it was gone in the tumor cells that grew out [that is, the daughter cells from that transplanted tumor]. And it turns out, that was the neo-antigen that was seen by the immune system. That allowed the tumor to be rejected spontaneously. "And eventually this evolved into the whole idea of, 'Oh, well, this is a pretty good idea, because it turns out that the T cells that get activated by the checkpoint antibodies like anti-PD-1 and anti-CTLA-4, those T cells are actually against these tumor-specific neo-antigens.'"

29. Gavin P. Dunn, Lloyd J. Old, and Robert D. Schreiber, "The Three Es of Cancer Immunoediting," *Annual Review of Immunology*, 2004, 22:329–360.

30. Jim Allison had helped find those checkpoints, developed inhibitors against them, and was in the process of trying to get those drugs into the clinic, to see if they worked on humans as an immunotherapy against cancer.

31. Dunn et al., "The Three Es of Cancer Immunoediting."

32. Daniel S. Chen, Ira Mellman, "Oncology Meets Immunology: The Cancer-immunity Cycle." *Immunity*, volume 39, issue 1:July 25, 2013, 1–10.

33. MDX-101 was developed in transgenic mice at Mederex by a team led by Alan Korman.

34. The anti-CTLA-4 antibody (MDX-010) was a human immunoglobulin antibody derived from transgenic mice having human genes. This antibody had been shown to bind to CTLA-4 expressed on the surface of human T cells and to inhibit the binding of CTLA-4 to its ligand (B7 molecules, expressed on antigen-presenting cells).

35. "Before clinical use, MDX-010 anti-CTLA-4 Ab [antibody] underwent extensive evaluation in cynomologus [macaque] monkeys and did not cause any notable clinical or pathological toxicity at repeated i.v. doses from 3 mg/kg to 30 mg/kg in acute and chronic toxicology studies (unpublished data from Medarex)." Giao Q. Phan et al., "Cancer Regression and Autoimmunity Induced by Cytotoxic T Lymphocyte-Associated Antigen 4 Blockade in Patients with Metastatic Melanoma," *Proceedings of the National Academy of Sciences of the United States of America*, 2003, 100:8372–8377, doi:10.1073/pnas.1533209100, http://www.pnas.org/content/100/14/8372.full.

36. Phan et al., "Cancer Regression and Autoimmunity."

37. "All of these patients had undergone surgery to remove the primary tumors, almost half had tried chemotherapy, and almost 80% of these patients had already undergone some form of immunotherapy; that included IFN-α (patients 2, 5–8, 10, 12, and 13), low-dose IL-2 (patients 2, 5, and 13), high-dose IL-2 (patients 4, 7, and 8), whole-cell melanoma vaccines (patients 1, 2, and 6), NY-ESO-1 peptide vaccine (patients 4 and 5), or granulocyte-macrophage colony-stimulating factor (patient 9)." Ibid.

38. The most dramatic of these patient stories was a woman who had barely passed the physical requirements for participating in the study. She had tumors collapsing one of her lungs and even more filling her liver, and all previous measures to stop the disease had failed. After a small, single test dose of the anti-CTLA-4 antibody she had gone into rapid remission, and by the time she left the study completely she had no evidence of the disease; the tumors had all disappeared. This complete response would turn out to be durable as well; fifteen years later this patient is still cancer-free. Dr. Antoni Ribas was clinical lead for this groundbreaking clinical study and a recognized leader in making anti-CTLA-4 a success.

39. Phan et al., "Cancer Regression and Autoimmunity."

40. Only fourteen had been able to complete both phases of the trial.

41. "We'd tested against a lot of tumors in mouse models, and eventually we realized that tumors that have a lot of mutations, and therefore a lot of neo-antigens, respond well," Allison says. "Those that don't, don't."

42. It originates on the part of the body (skin) most exposed to UV from sunlight and other outside carcinogens, resulting in tumors marked by a high number of mutations.

43. Those small mutational changes were often enough to allow melanoma to get "lucky" and escape whatever cancer drugs were being thrown at it. One drug

would work and kill most of the cancer cells, but the remaining cells continued to mutate, and if one of those mutations happened to be resistant to the
drug, that cell survived, and continued to divide. The new, drug-immune cancer would come roaring back, and the process would begin again with another,
less effective therapy. The patients who would be enrolled in a clinical study for
an experimental treatment had already tried all the available treatments. Their
melanoma had beaten them all.

44. He'd seen the relief of the lucky fraction of metastatic melanoma patients who
responded to chemotherapy, and then watched only a few months later as the
cancer came roaring back, mutated and stronger than ever.

45. One reason is that he entered the field, incredibly, as a teenager, mentored by
giants in the immunotherapy field. The other is that he works constantly, a habit
he was raised with. He grew up with a father who was a Teamsters official and
taught at New York City community colleges at night, and a mother who somehow survived a life spent as a New York City elementary teacher. (This hard
work trait is common to good doctors everywhere, but somehow nearly universal among the immune oncologists I spoke to, many of whom end up married to
lab partners or other immune oncologists, so they never have to talk about anything less important or interesting. Others, like Steve Rosenberg at the NIH,
seem to live on burned coffee and treat the lab as home.)

The traits combined while Wolchok was still in high school and took a
summer job in a Cornell immunology lab, working directly with patients and
vaccines. It doesn't get much more direct than that—except that it did the following year, when he went to college and met Lloyd Old. Old recognized the
interest and potential in the wunderkind and in 1984 introduced him to Alan
Houghton, the freshly tenured chair of immunology at the vaunted Memorial
Sloan Kettering Cancer Center. Cancer immunology wasn't exactly the obvious
choice for a Staten Island kid paying student loans to become a doctor—there
were easier ways to make good. But Wolchok is passionate and compassionate and intellectually driven. Like his friend Dan Chen on the West Coast, he
could think of nothing more interesting or useful than combining an MD with
a PhD and translating lab work to real people who needed it.

With Old and Houghton behind him, his path was set, if he wanted it, and
once again, Jedd Wolchok, as he puts it, "raised his hand for that." That summer he was helping out with a phase 1 clinical study using an antibody targeting
melanoma, in the lab by night, with the patients by day, living in the intersection of science and medicine, with living proof, in the form of his "anecdotal"

responders. The immune oncology worked and was real. It all clicked, and the course of his life was set when he was just nineteen years old. And while it's difficult to imagine a cancer immunology origin story that tops those name checks, now here he was more than a decade later, partnered with Jim Allison on the clinical study that would change everything.

46. Immune oncology wasn't the safest career path, especially if you had the education and the pedigreed training to work anywhere, including on therapies that were actually showing progress. That was why he knew Dan Chen—essentially, how could he not know another Gen-X MD-PhD oncologist specializing in melanoma and intellectually dedicated to beating it through immunology? People like that—Pardoll, Hodi, Butterfield, Hoos—were scarce compared to the chemotherapy-focused oncologists. That was the normal way—studying approaches to attack the tumor with drugs, rather than trying to figure out how to unleash the immune system to do the job. It wasn't the intellectual position you would expect from an otherwise promising young kid from Staten Island with decades of education, training, and student loans under his belt.

47. For Wolchok, one of those glimmers came in what seemed to most a failure: the test of interleukin-2, or IL-2, a cytokine or immune hormone. IL-2 was heralded as the success of the age, the game changer—it was seen, at the time, as the potential breakthrough. But once it was cloned in sufficient quantity to begin large-scale, systematic testing on patients, it didn't work as predictably as had been hoped. Instead of a breakthrough, IL-2 was declared a failure, as was the quest for a means to use the immune system against cancer with it. The experience set the public face of immunotherapy back decades.

Looking back at the data from those IL-2 trials, 3 to 5 percent of patients had positive responses to the immune hormone injections. But the patients who did respond were patients with melanoma and kidney cancer. And that group proved to be a small but reproducible number. That reproducible data showed researchers that IL-2 led to the growth and differentiation of T cells. The exact biological mechanism of how it did so wasn't fully understood, and nobody yet knew that cancer had other tricks that blocked immune response, such as CTLA-4 (which blocks general immune mustering of T-cells) and PD-L1 (which tumors express to put the brake on the T-cell at the moment it's targeted the tumor for attack).

And so the public opinion of IL-2 failure as the breakthrough against cancer was seen rather differently by some scientific researchers. It stomped on the spirits of many cancer immunotherapy faithful, even as it increased their faith in the concept of cancer immunotherapy. There, in black and white, were trials that showed that a drug (a cytokine, in this case) could in some patients lead to durable and profound immune responses—and long-term regression of cancer. The numbers were low, but they were reproducible. It wasn't a success, from a drug perspective, but for Wolchok and a handful of others, that was the first glimmer that proper modulation of the immune system could lead to durable control of tumor growth. It was proof of concept.

"People said, 'Oh, [IL-2] is too toxic. It doesn't work in many people'—and all of that is true," Wolchok says. "But it did show us that under certain circumstances—circumstances which we didn't completely understand at the time—the immune system could recognize cancer. And the immune system could control it. So, we had these glimmers. You start to put these different pieces together—you had the small reproducible successes of IL-2, you had the mouse models, you had the veterinary oncology—glimmers and glimpses. But [to the larger scientific community] this was considered extremely misty. And there were a lot of points to connect."

What was missing was some hard science—basic research into the misty workings of the immune observations they'd been making. And more than anything else, what they needed at that point was the missing puzzle piece, or maybe a complex of pieces that connected all the glimpses and glimmers, and turned the anecdotes into science. Chemotherapists and the majority of oncologists thought that this hope without proof was a bit flaky and rather misguided. But for the immune oncology faithful, this was exactly what a complex biological pursuit would look like. Yes, immunotherapy didn't work, but that didn't mean it wouldn't or couldn't. It was like they were trying to start a car; they'd seen other cars running fine, they'd observed that the engine sputtered but wasn't reliably turning over. Yes, definitely, they couldn't figure out why it wasn't running. But they still believed it was a car.

They believed that something real and specific was preventing the engine from turning over. They believed that the parts of the car they knew about—the ignition, the engine—all existed and were crucial to making the car go. They'd observed it running occasionally, seen other cars that ran, and were now finally working out all the mechanics of the system—the key that needed to fit, the gas pedal and lines, aspects of the engine, requirements of fuel and temperature and flammable gasses. Given all their understanding, they still couldn't make the car go. Sometimes, even when they could make the engine run, they still couldn't make it go forward. To immunotherapy researchers, that meant that there had to be another necessary part yet to discover, a mechanism they'd failed to understand. And they believed that if they kept trying, sooner or later they'd discover that thing. It would be the breakthrough that explained the problem. It was hopeful and inspiring, and then, like most immunotherapy at the time, frustrating. "You could see it working in the mouse models. But the challenge was to take what you'd seen in a 20-gram lab mouse, inbred and eugenically identical to every other mouse you're studying, and then try and translate that to a 70-kg outbred human," he says. And from there, they could make the car go, and start to work on better and different cars.

48. "We believed that under certain circumstances, you can develop protection from cancer by the immune system," Wolchok says.

49. At the time Wolchok was trying to develop cancer vaccines, and in an effort to overcome some regulatory hurdles, they started with clinical trials in dogs—not sad lab animals but people's pets, some he knew personally. Just like people, most of the dogs were outbred genetically, which is to say, they were mutts. And just like people, these pets had developed melanoma by an unfortunate interaction between genes and environment. In the dogs, his vaccines worked—"We showed that we could in fact change the life expectancy of a dog with metastatic melanoma by vaccinating it," he says. "The results were some happy dogs and dog owners and the first approved cancer vaccine (though it's only for dogs)." But the larger point was that Wolchok had seen an immuno-oncology treatment work with his own eyes. The vaccines wouldn't be the same, but the theory was identical: The immune system could be helped to recognize a mutated cell and kill cancer.

50. One of the most important conclusions was that the response rate they'd agreed to and set up is not in fact a terribly biologically relevant endpoint for testing the effectiveness of CTLA-4 blockade in cancer patients.

51. And he presents a very different vibe from the long-haired true believers that had dominated the field for a generation.

52. His triple-axel-and-a-somersault career had arced with mathematical precision, and now he'd stuck the landing. "Oh ya, it was good," Hoos says succinctly. "The right things came together at the right time and after many years of failure and disbelief." There was only one problem: That study he'd inherited wasn't going to work.

53. As a standard it's called "Response Evaluation Criteria in Solid Tumors," or RECIST. RECIST is the set of rules governing clinical trials, and how to measure a patient's tumor changes.

54. The problem with cancer isn't the cancer you have, it's the progression.

55. The unique language of cancer carries our collective history with the disease. Just as the name *cancer* conveys the image of a crab-like tumor, most probably a sarcoma, progressed to the point of bursting from the skin.

56. PFS describes a game of inches. It assumes the worst and counts small blessings. It also engenders a very specific way of thinking about the disease—in terms of the mechanisms of chemotherapy or radiation or small molecules that starve the tumor. These were the methods science was already familiar with. Over time, they became a habit and an intellectual blind spot.

57. The effects of radiation or chemotherapy upon a tumor target work in essentially the same way. The radiation sends tiny particles from a decaying isotope punching through the cells where it has been implanted like a miniature grenade. Chemotherapy essentially poisons them. The main force of the radiation or chemical attack is the tumor itself.

58. The primary investigators in charge of the Ipi studies were Dr. Steve Hodi (now at Dana-Farber Cancer Institute in Boston), Jedd Wolchok, Jeff Weber at USC, Khan Hanumui in Vienna, Steve O'Day at the Angeles Clinic in Santa Monica, together with Omid Hamid in Los Angeles and Dr. Ribas. They knew the numbers were bad.

59. "Mr. Homer" is not his real name; he is the patient referred to as case #2 in Yvonne M. Saenger and Jedd D. Wolchok, "The Heterogeneity of the Kinetics of Response to Ipilimumab in Metastatic Melanoma: Patient Cases," *Cancer Immunity*, 2008, 8:1, PMCID: PMC2935787; PMID: 18198818 (published online January 17, 2008).

60. Sharon Belvin has now been totally cancer-free for more than twelve years. "We looked at her CAT scan," Wolchok remembers. "The cancer was gone—all of it, gone—that does something to you."

61. Saenger and Wolchok, "Heterogeneity of the Kinetics of Response."

Six: Tempting Fate

1. The history of interferon is fascinating and popularly misunderstood (a very good treatment of the subject may be found in Stephen S. Hall's *A Commotion in the Blood*). It's also a case study in the difference between the public perception of what constitutes scientific breakthrough and true scientific advancement—a difference that can be summarized by the word *medicine*.

   Like most science stories, the story of interferon starts with a mysterious observation, a phenomenon that had first been observed, or at least described and published in the medical literature, in 1937, when two British scientists noted that monkeys infected by a virus (in this case the Rift Valley fever virus) were somehow resistant to infection by the yellow fever virus. The concept of inoculation and vaccines was already familiar, but this was something new; the two viruses did not appear to be related, and in fact the Rift Valley virus was fairly weak, unlike yellow fever, which would have killed them.

The observation played out time and again in various cells and animals; exposure to one virus, usually a weak, nonfatal sort, somehow locked out infection by a second, even fatal virus. Because infection by the first virus interfered with the ability of the second virus to get a foothold in the host, the phenomenon was called "interference."

The name given to the phenomenon says much about the time, and the way the mechanism was perceived to have worked, as if the first virus acted as a signal jammer for the second, the way a large radio tower broadcasting at fifty thousand watts crowds out smaller radio stations on the dial. In effect, the first virus seemed to create a sort of invisible force field repelling and protecting against infection by the second virus—perhaps in the form of some generated chemical shield, or perhaps caused when the first virus consumed all the resources that viruses need, making a second infection impossible, or perhaps—well, the literature and lunch tables of famous medical research centers across the world were littered with what-ifs. The mechanism of it wasn't understood at all, but its mystery seemed to be wrapped up in the secrets of the immune system and the biology of molecules and cells. Until you understood them, it was a nifty trick, if only a trick, but one with obvious potential practical application, and thus reproduced, in the 1940s and 1950s, in dozens of labs and causing an entire generation of scientists to look at viruses and virology as the most interesting and important topic on the block.

In 1956, the center for that research was in a series of unassuming buildings at the National Institute for Medical Research, a place located on a rise called Mill Hill, north of London. The United Nations' World Health Organization labs were also headquartered there, with an early-warning World Influenza Center overseen by none other than C. H. Andrewes, the legendary virus researcher who had discovered the virus that causes influenza in the 1930s.

In June 1956, Andrewes's lab was joined by a thirty-one-year-old biologist fresh off the train and ferry from Switzerland named Jean Lindenmann. Lindenmann's employment in Andrewes's lab was related to polio; there had been successes with the first generation of polio vaccines, and Andrewes hoped to improve upon it, but he needed great supplies of the virus to work it out with. Lindenmann was to attempt to grow the viruses in rabbit kidneys. He failed, but ended up collaborating on another experiment that was far more successful.

While most of the real scientific work happens inside the laboratory, story after story demonstrates that the lunchroom, where scientists speak about their findings and passions and exchange ideas with members of other laboratories, is the true breeding ground of invention. Just so were the famously overcrowded lunch tables of the Mill Hill canteen. It was here, hunched over his hurried lunch, that Lindenmann happened to find himself talking about his fascinating interference phenomenon with a charming and accomplished virologist named Alick Isaacs. Lindenmann was surprised to find that Isaacs had become fascinated with the magic of this interference thing, too. Isaacs was remarkably enthusiastic about it—not only because he happened to be a manic depressive on an upswing, which he was. As it happens, the young scientist had already gone ahead and conducted several experiments on the thing.

Lindenmann had an experiment in mind, too, one he believed would help answer a critical question: Did a virus need to actually enter a cell in order to imbue it with the magic "interference"? He hoped the answer could be made clear—visible, in fact—using the powerful new tool the lab had its disposal: the electron microscope. At the very least, the experiment would bring them a step closer to understanding whether the interference force field was something that occurred *around* the cell (a phenomenon on the surface or immediate surroundings), or whether the switch had to be turned on from *inside* the cell. Isaacs was game.

2. Chen is well trained as both an immunologist and an oncologist. Nothing could be more interesting to him than the interface between his two chosen fields. But he wasn't ready to go all-in on joining the ranks of the confirmed true believers of the immuno-oncology ranks, either. Maybe it was because he was raised to be practical, to keep his feet on the ground while moving toward a career, even if he occasionally glances up at the stars.

3. Chen and family are a fascinating bunch, scientific and yet theatrical, though cancer immunologists tend to stray from the nerdy and narrow scientific stereotype. Dan Chen loves music, science, collecting Pappy Van Winkle and other whiskeys, and Halloween houses, and he and Deb have built a family of three smart, kind, and talented kids—it's all a bit intimidating on a visit. On top of that, one has to remember that his day job is literally curing cancer. "I love what I do," Chen explains. "I live for this [cancer immunotherapy research] and

there's no gap between my personal life and my work life." It's a sentiment I heard often among those dedicated to the field.

4. Chen was a Howard Hughes Medical Institute associate at Stanford; he completed his internal medicine residency and medical oncology fellowship there in 2003.

5. Chen had contacted Dr. Weber again to see about getting him into a different, newer vaccine trial, but his bout with the previous vaccine made him ineligible for such a study. If Brad was going to stand a chance he'd need something else—something available now.

6. Dr. Chen: "It would require the ability to quickly and economically sequence the entire genome of both the patient and that patient's cancer, a computer powerful enough to run bioinformatics on all that data and determine the best antigen target that would get a strong response to the tumor without creating a response against the patient, and the ability to turn that into a custom vaccine. We can do all of that now."

7. Dr. Jeffrey S. Weber, MD, melanoma and immunotherapy oncologist, is currently at NYU Langone.

8. Working so closely with a patient like Brad was one of the reasons Dan Chen was perfectly happy at Stanford. He had his lab and his oncology practice. He didn't have a specific interest in leaving that for what they call "industry"— working for a company, rather than a university research facility. His interaction with patients nourished him, and the academic setting was what he'd aspired to. What he wanted most was what his patients wanted—answers. Hope. New solutions. And during the time he was thinking that, an offer came for him in 2006 to join the team at the local biotech.

His first impulse was to ignore the solicitation. There was virtue in the university research setting, and he worried that there might be something mercenary in leaving that for a for-profit company. Besides, he had worked hard and built a good career in academia. He had a sharp young team assembled in the lab, he was pleased with his research, and he was publishing well and often. He was climbing the ladder. The university setting offered built-in stability, the sort his academic parents worked hard to accomplish for themselves and had always imagined for Dan, too. So he wasn't inclined to leave that and not at all inclined to leave his patients, some of whom he'd been seeing for years. But when the call came, he didn't see the harm in at least hearing what they had to say, and maybe having a meeting.

9. From: Daniel Chen < Subject: Melanoma
Date: Thursday, February 18, 2010, 5:30 PM

Hi Brad,

I received your message, and I'm certainly disappointed, as I'm sure you are. However, I am happy to hear that the recurrence appears to be occurring in the same place it was seen last.

-Did you ever get radiation treatment at that site?

-Have you had your tumor analyzed for the V600E bRAF mutation?

-Have you contacted Don Morton regarding surgical resection?

-Would you consider IL-2 treatment at this point vs. clinical studies?

10. Leslie A. Pray, "Gleevec: The Breakthrough in Cancer Treatment," *Nature Education*, 2008, 1:37.

11. The so-called Philadelphia chromosome, or BCR-ABL, a fusion of two swapped genes found to be present in 95 percent of patients with a specific type of white blood cell cancer, chronic myelogenous leukemia (CML). The work, begun in the 1960s, was the first time a connection between a genetic condition and a predilection for cancer was realized. The drug remains transformative for that subset of patients.

12. "I wouldn't say Ira was covert about it, but he wasn't exactly open, either," Chen says. Cancer immunotherapists were a special little minority. The larger body of cancer biologists looked at that special little group as being . . . different. Bad different. "Passionate," is how Dan puts it, "but maybe overly passionate." Which is how everyone else says "crazy." The feeling seemed to be that they were a group for whom belief had clouded their scientific objectivity; they believed, and so they did not see. Declaring your faith in the promise of cancer immunotherapy at a drug company meeting was a surefire way to have your ideas discounted. "They thought it's all bogus," Dan says. "I think a lot of cancer biologists recognized the promise of the field. But a lot of them, I think, felt that the biology just wasn't there, they just didn't believe in it. It wasn't the future. Especially since they'd discovered the oncogene for melanoma—that was the future, targeted therapy."

The room was polarized between the cancer biologists and the cancer immunologists. In the middle was Scheller. The cancer biologists were excited about the identification of the oncogene that drives the mutations that turn a cell into melanoma. In the room, if it was a vote, it's 50 to 80 percent certain that they'd want to develop a drug to target the transcription of the oncogene into melanoma.

13. Steinman died of cancer in September 2011, only a day before the committee quietly notified the winners.

14. Ira Mellman remembers they'd argue about it, but they'd never convince anyone.

"The challenge of immunotherapy is that it was a promise for a hundred years, OK? With breakthroughs always twenty years away," Mellman says. "So the idea has been around probably for the better part of the century now, if not longer, that you can activate an individual's immune system to combat cancer. But when that first hit—it hit about the same time as surgery did, and when radiation therapy did, so it was sort of pushed off in the corner, in part because so little was known about the immune system at that time, and in part because the work sucked, from a scientific perspective. And that motif remained for decades!

"The cancer biologists had the oncogene to point at. The cancer immunologists had some exciting new papers coming from cancer immunotherapy research, and some startling new data. The data was empirical, but what it meant was still open to interpretation and bias," Mellman says. "You don't argue the facts, but you argue the interpretation." But for every study which suggested one truth about getting the immune system to recognize cancer, there was always another, seemingly just as credible, which suggested the opposite. The cell biologists pointed to the one, while the immunologists pointed to the other. There was no lack of data, numbers, studies, or the usual mouse models; they'd all seen them. And there were problems with mouse models. "They're rarely predictive of humans" to begin with, Mellman explains. "The mouse models were always crummy." Maybe one out of five worked. But the one that worked, worked really well. At the end of the day, the cancer immunotherapy cynics had the trump card, which Mellman paraphrases as "You don't even know how any of this works." And the worst part was, they were right. He adds, "The underlying mechanisms, the biology—if you don't understand that, how can you tell us you really understand these new findings?" And the truth was, they couldn't. Nobody understood the complex biology. And it was nearly impossible to make a solid scientific argument without the science to back it.

15. To figure out which T cell genes *didn't* have anything to with the self-destruct signal; the genes led to the receptor.

16. Y. Ishida et al., "Induced Expression of PD-1, a Novel Member of the Immunoglobulin Gene Superfamily, upon Programmed Cell Death," *EMBO Journal*, 1992, 11:3887–3895, PMCID: PMC556898.

17. He is now codirector of the cancer immunology program at Yale Cancer Center in New Haven, Connecticut.

18. B7-H1, a third member of the B7 family, costimulates T-cell proliferation and interleukin-10 secretion. See H. Dong et al., "B7-H1, a third member of the B7

family, co-stimulates T-cell proliferation and interleukin-10 secretion," *Nature Medicine*, 1999.

19. Lieping Chen had cloned the human PD-L1 gene and, he says, attempted to convince a company to produce a commercial antibody against it in 2001, without success.

20. The testing was done at NIH, initiated by Dr. Suzanne Topalian.

21. Dan Chen had to look no further than his own childhood dining room table. He could still see his father there, a physicist passionate about his work on equations to realize the dream of fusion.

  "And it's the same deal, right? A passionate group of scientists believe in fusion as the future of energy, and it was always twenty years away," he says. "And now here we are, forty years later, and there's still this passionate group, and it's still twenty years away. And I think the worry was, you know, we've got the small steps. We knew the biology was there. But was it always going to be twenty years away when we had something that was really useful for patients? And so none of us could really say when the actual breakthrough that would make this useful for the majority of patients was going to happen."

22. One of the arguments Dan made for immunotherapy was its value proposition, which he summarized through a story about a night he and Deb were having dinner at a friend's house. Dan hung out in her kitchen drinking wine while his hostess chopped salad. As Dan recalls, he started telling his hostess about his work and the progress they were making against cancer. He was wrapped up in it, and excited. The trials had come back and shown the drug would help cancer patients live longer.

  "Oh, that's great!" she said. "How much longer?"

  Dan remembers telling her. Sometimes it was a few months.

  "That's it?" she said. "I thought you were curing cancer!"

  He heard himself explaining that, well, cancer was really hard, and these numbers are averages, and—but he was justifying what he himself was frustrated about. Yes, he was in drug development now, that was exciting. Yes, they were making progress, the same incremental progress, the weeks and months that add up to years. That had been the cancer therapy story for at least a generation, maybe two. He was chipping away. But he wasn't breaking through. None of them were.

  When the first cancer research labs had been set up in the early part of the last century, the cure was the goal. They believed it was possible. And why shouldn't it be? Other diseases had been cured through directed study and good science and a great deal of money. New technology was slowly cutting a swath through the forest of plague and pestilence that had hectored mankind for eons. The best minds were in the field. And one hundred years on, they'd made definite improvements for cancer patients. But they hadn't found a cure.

23. At the time it was known the molecule they'd found was a protein expressed on tumor cells, but its connection to a receptor on the T cell, or the notion that interaction with that receptor would down-regulate T cell response, was not even guessed at. Instead the protein was seen as a potential target on tumors, a sort of molecular bull's-eye that you could target with a corresponding antibody. A more typical drug development approach to cancer at that point was to then attach that antibody to some sort of poison you wanted delivered to the cancer cell. That was the process of drug development the team was headed toward when the immunologists in the room steered them off course.

24. "Everyone was willing to accept that PD-L1 might work in melanoma and kidney cancer," Chen explained. These highly mutated cancers (especially melanoma) had also seen promising results with anti-CLTA-4. "But, even internally, the skeptics would say, 'I'll believe it when it works in lung cancer.'"

25. Another type of T cell, called a T regulatory cell, or T reg. The role of these cells is still being explored, but they are increasingly understood to be critical players in the checks and balances of immune response; they are, in a sense, cells always looking to call a truce to immune battle. It has not yet been definitely determined which is the more important influence, stimulation of T cell response or down-regulation of the T regs; in all likelihood both may end up being important.

26. "I was close with Brad," Chen says. "That friendship made the highs more extreme and personal. And it made the lows personal, too."

Seven: The Chimera

1. Dr. Michel Sadelain, comments to the author.
2. Based on the work of Tak Mak and others.
3. "Zelig made the receptor, I put it into T cells," Hwu says. They started using patient's T cells against melanoma, and then retargeting those TILs for ovarian, colon, and breast cancers. The ovarian cancer retargeting worked best of the three—the reengineered T cell recognized the antigens of the IGROV ovarian cancer cell line. "The first time I got that to work I was so elated," Hwu remembers. But successful retargeting was only one part of engineering a successful cancer-killing machine. Such a cell also needed to be able to stay alive in the

body, replicate into a clone army, and successfully and selectively kill the targeted cancer. In these regards, the cells reengineered at the NCI did not work.

4. This CAR-T was a far cry from the model T-body of 1985, a sleek, complex killing machine. "The first-generation CAR could, when put into a T cell, recognize the target molecule and kill the cell," Sadelain explains. But they also must proliferate—they must grow and clonally expand. They must also remain functional T cells and retain that function over time. That required further modifications. Sadelain's innovation was to introduce a costimulatory signal and produce what he calls a "second-generation CAR" that recognizes a target, expands clonally, and retains its other T cell functionality. Such a cell is "a living drug," with a life span as long as that of the patient it lives in. This work was based out of his lab at Memorial Sloan Kettering Cancer Center, where Sadelain is the founding director of the Center for Cell Engineering and head of the Gene Transfer and Gene Expression Laboratory.

In 2013 Sadelain formed a company called Juno Therapeutics to exploit the new CAR-T technology, together with his wife and coinvestigator, Isabelle Rivière, and partners Michael Jensen, Stan Riddell, Renier Brentjens, and Fred Hutchinson Cancer Research Center immunologist (and Jim Allison's Deadhead pal) and true-believer immunologist Phil Greenberg. The race was on to turn the potential killing machine into a more effective weapon against cancer.

5. Dr. June had based his CAR design on a sample he had requested from Dr. Dario Campana, then of St. Jude Children's Research Hospital, after hearing a presentation by Dr. Campana at a 2003 conference.

6. The CD19 protein target selected by Sadelain was an essential aspect of the success of CAR-T and, in essence, he says, launched the field. "CD19 was known, but it was not a star when I selected it," he explains. The criteria for a good molecular target for a CAR to recognize was that it be unique to cancer; if the antigen was found on cancer cells but also expressed in normal body cells, the CAR-T would attack both the cancer and the host. CD19 was a good choice, because it was an antigen that was heavily on the surface of certain cancers such as lymphoma. It also is expressed by some B cells, but that was collateral damage that was survivable; physicians have long experience keeping patients alive without B cells. "Facing terminal cancer, losing your B cells isn't so bad," he explains.

In a 2003 paper in *Nature Medicine*, his group showed that you could collect T cells and introduce a retroviral vector that coded for a second-generation CAR that recognized and targeted CD19 in animal models (immunodeficient mice given human genes and human CAR-T cells). That proof of concept in a preclinical model would then need to be approved for testing in a clinical setting—and the decision to allow genetically engineering a killing machine that targeted human proteins to be tested in a human subject would need to be carefully considered by the Recombinant DNA Advisory Committee (RAC) as well as the FDA.

7. And his CAR was gassed up by also expressing a costimulatory protein (called 4-1BB) that was similar to CD28. The result, they hoped, was a CAR with a steering wheel pointing it where they wanted it to go, and enough fuel to keep the T cell going long enough to get there and finish the job.

8. In 1991, Arthur Weiss of the University of California, San Francisco, developed a chimeric antigen receptor (CAR) called CD4-zeta as a means of studying the activation of T cells. See Jeff Akst, "Commander of an Immune Flotilla," *Scientist*, April 2014.

9. GVAX was based on work that combined the cutting edge of gene therapy with that of immunotherapy. It was focused toward what was seen as the most promising direction for cancer immunotherapy at the time: the development of a cancer vaccine.

The treatment took a piece of a patient's tumor, altered the tumor cells' genes so that they would express a cytokine (called granulocyte-macrophage colony-stimulating factor, or GM-CSF, which had recently been shown to be involved in getting dendritic cells to present tumor antigen to T cells, work done by Ralph Steinman), and then reinject the modified tumor as a sort of double-duty vaccine, altering the immune system to the tumor while producing cytokines that stimulated the response. That was the theory anyway, but like all cancer vaccine trials during the 1990s and early 2000s, it failed, and the work was essentially shelved in 2008. The reason for such failure isn't certain, of course, but much more is now known about the biology of the immune system, cancer, and the immunosuppressive tumor microenvironment, including the expression of PD-L1.

This was a fascinating chapter in the immunotherapy story, involving MD-PhDs now recognizable as a veritable who's who of cancer immunology, including Glenn Dranoff, Richard Mulligan, Drew Pardoll, Elizabeth Jaffee, and others. Each of these researchers and scientists deserves a chapter in this book, and nearly all of them are currently doing important work that will surely write the next one. (Elizabeth Jaffee, for example, is working with GVAX in combination with an anti-PD-1 checkpoint inhibitor, nivolumab, in pancreatic cancer. Another combination approach is being evaluated by Aduro Biotech partnered with Novartis, where Dranoff heads oncology drug development, and Pardoll is codirector of cancer immunology and Hematopoiesis Program Professor of Oncology at John Hopkins in Baltimore.)

That original 1993 academic paper that laid out the scientific foundation and therapeutic claims for GVAX can be found here: Glenn Dranoff et al., "Vaccination with Irradiated Tumor Cells Engineered to Secrete Murine Granulocyte-Macrophage Colony-Stimulating Factor Stimulates Potent, Specific, and Long-Lasting Anti-Tumor Immunity," *Proceedings of the National Academy of Sciences of the United States of America*, 1993, 90:3539–3543.

Incidentally, Dan Chen had presented this paper for his journal club while still a medical school student, and it had provided a spark of interest that helped shape his career. Years later he would marvel that those researchers were now his peers in the small world of immunotherapy. Or, at least, a world that was small until the breakthrough.

Several of these fascinating and critical players in the immuno-oncology world were also interviewed by Neil Canavan, a writer and researcher with the biotech venture capital firm the Trout Group, for his book *A Cure Within: Scientists Unleashing the Immune System to Kill Cancer* (see "Further Reading," following the appendixes).

10. June still remains dedicated to research for ovarian cancer specifically, as well as for the blood cancers currently targeted by CAR-T therapy.

11. Children who undergo chemotherapy and radiation for the treatment of blood cancers are often cured, but they suffer as adults do not—one of the reasons that children with leukemia are now eager to skip those therapies and go directly to CAR-T. Learn more at EmilyWhiteheadFoundation.org.

12. The Whiteheads had originally sought a second opinion at the Children's Hospital of Philadelphia and wanted to pursue the CAR-T therapy, but the FDA had not yet approved the therapy for pediatric patients. Pediatric therapies get a more rigorous and thus slower vetting than adult therapies, which physicians like June find frustrating, especially when patients' lives depend upon them.

13. Because viruses are agents of infection that cannot reproduce on their own, scientists are not in total agreement as to whether viruses merit a branch on the tree of "life," as we define it; they are mobile molecular assemblages that to some more closely resemble tiny organic machines than living creatures.

14. The scene, as reported by several outlets at the time, was remarkable and, like all children's cancer wards, heartbreaking. Emily lay on the hospital bed in a sparkly purple dress, bald and browless from the failed chemotherapy, a

pressure cuff around her thin arm. The feeding tube snaking up around her ear and into her nose was held fast by pediatric tape, purple to match the dress.

15. James N. Kochenderfer et al., "Chemotherapy-Refractory Diffuse Large B-Cell Lymphoma and Indolent B-Cell Malignancies Can Be Effectively Treated with Autologous T Cells Expressing an Anti-CD19 Chimeric Antigen Receptor," *Journal of Clinical Oncology*, 2015, 33:540–549.

16. Dr. Grupp is both Children's Hospital oncologist and the principal investigator of the CART-19 trial in children. See Jochen Buechner et al., "Global Registration Trial of Efficacy and Safety of CTL019 in Pediatric and Young Adult Patients with Relapsed/Refractory (R/R) Acute Lymphoblastic Leukemia (ALL): Update to the Interim Analysis," *Clinical Lymphoma, Myeloma & Leukemia*, 2017, 17(Suppl. 2):S263–S264.

17. We now have strong evidence suggesting that it is not the engineered T cells themselves that release the IL-6, but rather it is the macrophages (blob-like elements of the innate immune system) that surround the attack upon cancer and do the cytokine releasing. In June 2018 Dr. Sadelain's team at Memorial Sloan Kettering Institute released a letter in the journal *Nature Medicine* detailing this finding, as discovered via their CRS mouse models. One hope from this finding is that they will be able to identify the specific chain of molecular events in the cytokine cascade and block those that cause the dangerous symptoms, without interfering with those cytokines necessary for the coordinated immune attack. In doing so, CAR-T therapy may become less toxic and can be performed outside of a hospital setting.

Another hope is to take some of the variability out of CAR-T therapy, which, as a personalized medicine, varies in intensity from person to person. CAR-T is a unique drug in that it replicates itself in the body (unlike most drugs that are depleted by use), but not all T cells are equal. Those from a healthy immune-competent patient will replicate more prolifically than those of ill or older patients, or of those whose immune systems have been compromised by disease or chemotherapy. This makes dosing difficult for the clinician. Too few CAR-T cells results in an inadequate cancer-killing response; too many results in toxicity and CRS. See Theodoros Giavridis et al., "CAR T Cell–Induced Cytokine Release Syndrome Is Mediated by Macrophages and Abated by IL-1 Blockade," *Nature Medicine*, 2018, 24:731–738, doi:10.1038/s41591-018-0041-7.

18. As well as another cytokine-dampening drug, etanercept.

19. Tocilizumab is now co-indicated for CRS and is used on CAR-T patients.

20. James N. Kochenderfer et al., "Eradication of B-Lineage Cells and Regression of Lymphoma in a Patient Treated with Autologous T Cells Genetically Engineered to Recognize CD19," *Blood*, 2010, 116:4099–4102, doi:10.1182/blood-2010-04-281931.

21. In the *New York Times*, writer Andrew Pollack relayed a telling story about Steve Rosenberg as told by Arie Belldegrun. Prior to heading Kite Pharma, the National Cancer Institute's corporate partner in commercializing CAR-T, Belldegrun had been one of hundreds of former research fellows whom Dr. Rosenberg trained and mentored over his career. At the time, Belldegrun had been trying to recruit Rosenberg to join his company, an offer that would have certainly made Dr. Rosenberg a very rich man (in 2018 Belldegrun and his partner sold Kite for more than $11 billion).

    "He sits quietly, quietly, quietly," Belldegrun told Pollack, "and then he asks, 'Arie, why don't you ask me what I want to do?'

    "He said: 'Every day that I go to work, I'm as excited as a kid coming to a new place for the first time. If you ask me what I want to do, I want to die on this desk one day.'" (See Andrew Pollack, "Setting the Body's 'Serial Killiers' Loose on Cancer," *New York Times*, August 2, 2016.)

22. The generic name for this CAR-T is axicabtagene ciloleucel.

Eight: After the Gold Rush

1. Another early therapy from this proof-of-concept era, approved a year before Ipi, was a dendritic cell therapy developed by a company called Dendreon. The drug, sipuleucel-T, failed to become commercially viable.

2. A partial list of those those PD-1 blocking agents starts with pembrolizumab (trade name Keytruda, developed by Merck pharmaceuticals and approved in 2014) and nivolumab (trade name Opdivo, developed by Bristol-Myers Squibb and approved in 2015). The drugs targeting the tumor side of the handshake (anti-PD-L1) rolled out soon after. One was atezolizumab (trade name Tecentriq), developed by Genentech and Roche pharmaceuticals, which received its first FDA approvals in 2017) and durvalumab (trade name Imfinzi), manufactured by AstroZenica and MedImmune and approved in 2018.

3. A May 2018 letter to the *New England Journal of Medicine* reported on a subset of patients whose tumors had been shown to grow, rather than shrink, during a phase 2 clinical trial of the PD-1 checkpoint inhibitor nivolumab (Opdivo).

These patients had an aggressive and relatively rare form of cancer that affects the T cells, called adult T cell lymphoma-leukemia (ATLL). Lee Ratner et al., "Rapid Progression of Adult T-Cell Leukemia-Lymphoma After PD-1 Inhibitor Therapy," letter to the editor, *New England Journal of Medicine*, 2018, 378:1947-1948.

4. In the battle metaphor, blocking that checkpoint tells the army to grow and arm and prepare to attack. PD-1 / PD-L1 is a checkpoint that happens later, up close and personal, after the T cell army is already mobilized and ready.

5. Immunologists now categorize the interaction between an individual's immune system and their specific tumor into three broad classes: "hot, cold, or luke-warm." The categories are useful in describing how different tumor types, and different immune systems, present a range of dynamics which need to be addressed by different drugs or drug combinations.

"Hot" tumors are the ones most recognized by the T cells. Under the microscope you can see them massed at the tumor, and infiltrating inside the tumor ("tumor-infiltrating leukocytes"). They're there, and yet the T cells fail to complete the job and attack and kill the tumor. Also, these hot tumors may have various ways of "exhausting" T cells so that they cannot be "reactivated." (Remember that the immune system has a series of safeties and circuit breaker–like or timer-like elements to prevent every immune response from snowballing into a full autoimmune nightmare; even effective vaccines require a "booster shot" to reactivate T cell response.) As a result, they are present but too spent to attack. Many of these tumors tend to arise in parts of the body that are most exposed to stuff that causes cancer, like sunlight, smoke, or other carcinogens. They include skin cancers (melanoma), lung cancers (small-cell and non-small-cell carcinoma), and cancers arising in organs that deal with concentrated levels of the stuff that goes into our bodies, such as bladder, kidney, and colorectal. For DNA in the process of replicating itself, these carcinogens are like a con-stant bombardment. It would be like trying to write out a recipe while being pelted with golf balls; the odds are pretty good that you'll make plenty of mis-takes. In cells, these mistakes are mutations, and as you'd expect, the cancers that arise in these carcinogen-exposed organs are characterized by the highest number of "mistakes" in their DNA, and have some of the highest levels of mutations. Mutations (for these or other genetic reasons) make them highly visible to the immune system, which make them "hot." The fact that they're seen but not killed by the immune system means that something else is also happening, a trick that lets them survive despite being mutational peacocks.

In some cases, tumor expression of PD-L1 is one of those tricks, and as such, these tumors are the most likely to be expressing PD-L1, the secret handshake telling the immune system to pay no attention, despite all the antigens. As such they're also the tumor types most responsive to checkpoint inhibitors (anti-PD-1 or anti-PD-L1). Right now, these are the "lucky" tumor types, most likely to respond to the available immunotherapeutic drugs—and when they do respond, the responses can be profound. It's these tumors types that have oncologists willing to use the word *cure*.

An entirely different problem exists if the tumor is "cold." The immune system almost entirely fails to respond to these tumors. Under the microscope, you may believe that we have no immune system at all, which is why cold tumors are sometimes described as "immune deserts." These tumors are, for various reasons, less or not at all visible to T cells. Unlike their hot cousins, many—but not all—cold tumors are not highly mutated, and are not highly antigenic, meaning they don't present themselves as obvious to the immune system by presenting antigens that are clearly foreign. In this case, immune therapies that "warm" the tumor up and make it more visible (more antigenic) might be employed (such as targeting the tumor with a virus to mark it with more obviously foreign antigens). Cold tumors may also employ other tricks that prevent T cells from effectively recognizing them. Those may be aspects of the tumor microenvironment—the little world created by the tumor itself—where molecules (in various ways) disable or suppress the full immune response (which is also referred to as a "suppressive TME"). The majority of a tumor mass is not cancer, but components of the tumor microenvironment. And it's a tough neighborhood for a T cell to infiltrate.

Nature is conservative in that it doesn't tend to evolve complexity when simplicity is successful. To generalize, that's the reason most cold tumors don't respond to checkpoint inhibitors: They're the least likely sort of tumors to need a secret handshake like PD-L1 in order to survive and succeed. Their low mutation profile already makes them less visible to the immune system. With no checkpoint being taken advantage of by the tumor, inhibiting it will not change the situation.

And so as you'd expect, cold tumors respond poorly to the existing checkpoint inhibitors alone, and several types don't respond at all. To imagine why, it's useful to think of these tumors in terms of evolution. If a mutated cell is obvious to the immune system, the immune system sees it and kills it. The more mutated, the more obvious it is, and the less likely it is to survive and grow and become what we'd call cancer—unless it has also evolved a trick to compensate for its visibility. PD-L1 is one such trick. Cold tumors simply don't need such a trick.

A third tumor type is generally, if not helpfully, called "lukewarm." These tumors are seen by the immune system, the T cell army masses. But then, for some reason, the attack never happens. The T cells don't infiltrate, they don't destroy the tumor. Immunologists sometimes compare this to an army that has heard the battle call, massed at the castle, but cannot cross the moat. This category covers a wide variety of cancers and mutation types, and it would be incorrect to typify these cancers by any single factor. Unlike their description, it's not simply that these tumors successfully evade immune attack due to some averaging or combination of hot and cold attributes—though aspects of both may be true. It's most accurate to think of these tumors as having a unique profile of immune defense that allows them to survive and thrive without being totally invisible to the immune system. These tumors include some—but not all—glandular tumors. What matters is less often where these cancers arise than what typifies them.

In some cases, what makes them "lukewarm" is that despite being obvious, they exist in places that are difficult for immune cells to infiltrate. They may be typified by tumors with a tough outer layer that repels infiltrators. They may have evolved an almost fantastical outer line of defense. But generally, they are typified by a moderate expression of PD-L1, a moderate mutational load, a moderate antigen presentation, and often an immune-suppressive microenvironment that turns down immune response in the T cells at the gates. And there are some single therapies being tested to uniquely address these tumor types, but it's fair to say that various elements of hot and cold tumor approaches— including checkpoint inhibitors, therapies to warm the tumor by making it more immunogenic, and approaches to counteract suppressive elements in the tumor microenvironment—may all be considered in changing the situation to one where they are recognized, targeted, infiltrated, and destroyed by the immune system. Here too, various stages of the cancer immunity cycle are being targeted to get those T cells across the moat (to become tumor-infiltrating leukocytes), activated, and recharged.

Nine: It's Time

1. The footage is also now incorporated in the official music video of that Imagine Dragons song. See Jesse Robinson, "Imagine Dragons—for Tyler Robinson," You-Tube, October 27, 2011, https://www.youtube.com/watch?v=mqwx2fAVUMO.

2. The Tyler Robinson Foundation. Learn more at www.TRF.org.

3. Kim White wasn't in the band, but she had been a member of their extended Salt Lake City crew, a fellow Mormon who had first met the lead singer when they were both teenagers. Jeff Schwartz knew her as the tall, pretty, young blond girl—"very pretty, very blonde"—who sometimes came to the show with her husband. Jeff saw her again at her benefit.

4. Kim has written about her cancer journey, which appeared originally on the *Small Seed* and is available on the Deseret News website at https://www.deseretnews.com/article/865667682/Utah-mother-I-am-now-and-will-forever-be-grateful-I-was-diagnosed-with-cancer.html.

5. The page has been live since July 2014 and has raised $16,075 toward a new $50,000 goal.

6. Kim White: "My husband had reached out to Mac [Mac Reynolds, singer Dan Reynolds's brother and Imagine Dragons' manager] and he was like, 'of course.' Originally it was going to be the whole band, but it was so crazy with everyone's schedules, they were out of the country in the middle of a tour, so it was just Dan [Reynolds] who flew into Utah and did the benefit and they raised like $40,000 and he flew back the next morning." The original plan also involved making wristbands as well, rubber ones people could buy and wear in support, and they needed a name to brand it with. They settled on "KimCanKickIt," referring to the cancer, of course, as well as her love of soccer.

   If you want to follow more of Kim White's story she invites you to follow her at KimCanKickIt on Instagram.

7. She referred to her physician Dr. Boasberg as an "angel of a doctor."

8. Manufactured by Merck as Keytruda, most commonly used to treat melanoma. Following the 2013 announcement of results from the study at the Angeles Clinic and elsewhere Merck applied for breakthrough designation for the drug so it could be both made immediately available and fast-tracked to approval, and it received that designation in September 2014. In the summer of 2016 a clinical trial testing the checkpoint inhibitor for use against small-cell lung cancer was stopped; the drug was proving so effective that the company and the FDA wanted to provide it to everyone in the study, rather than deprive the control group (patients receiving a placebo or other treatment) of the opportunity. It gained formal FDA approval for this cancer in March 2017.

   Also in 2017 the drug was FDA approved for use against tumors that showed a specific mutation or genetic marker (microsatellite instability), making it the first drug to ever be approved for such an indication, and the first cancer drug to be approved for a genetic marker on a tumor rather than the organ in the body in which the mutated cell originated. This approval is hoped to be the first of many as tumor biomarkers are better classified and cancer cells genetically typed. If a tumor with a certain biomarker is known to respond to the drug, that is a far more efficient means of determining who would see benefit from it. This efficiency translates to patients trying to make crucial decisions about which therapy to choose, as well as to the drug companies responsible for conducting lengthy and expensive clinical trials for every type of cancer.

9. Clinical trials for drugs are specific to that drug being used as a therapy against specific cancers; while such drugs, once approved, can be used "off label" for unproven indications, it requires new clinical trials to access the safety and effectiveness of that therapy as opposed to others—an important distinction when patients often may not have time or health enough to try again.

10. Those original clinical trials were for melanoma; pembrolizumab was at that time called lambrolizumab. See Omid Hamid et al., "Safety and Tumor Responses with Lambrolizumab (Anti–PD-1) in Melanoma," *New England Journal of Medicine*, 2013, 369:134–144.

11. Kim's experience with her rare form of cancer has made her an inspiration and example for others with adrenal cortical carcinoma. "As far as I know there are only four other people with it on that drug right now," Kim told me. "Most people with it don't respond." But she did, and she immediately shared that on a group on Facebook for people with the disease, so they could try it, too. The anti-PD-1 drug did allow her immune system to successfully fight nearly all the lesions in her lungs, but it wasn't the end of her journey with cancer. Eventually her Salt Lake City oncologist told her he could get Keytruda for her there, so she didn't need to fly to LA every three weeks, which was inconvenient and expensive. "So that was great," she says. Even the parking was better. "We'd fly to LA and we'd rent a car and have to park it for like an hour [for the treatment], and the first time I realized, you have to pay for parking and it's like fifteen dollars and I was like, 'What the heck is that? I'm definitely not in Utah anymore!'" The first months after she started recovery were the first she got to spend time with her daughter as something other than a terminally ill person. "That was really important," she says. "She was only eighteen months old when this started, I'd never been anything else. So we spent those months backpacking, camping, really spending time."

But despite continued treatment, "for whatever reason, the Keytruda doesn't like my liver," she says. "It doesn't want to kill the cancer there." And months later, she discovered that the remaining lesion had continued to grow.

That required another surgery ("It was massive and they almost lost me, I almost died"), and the removal of 70 percent of her liver and a quarter of one of her lungs. "I spent a year recovering from that," she says.

And she still is. Kim isn't cured, but she's alive, enjoying life despite the daily injections to thin her blood, the regular chemo she still receives, and a near-constant litany of tests and upkeep against the disease. "I definitely know it's been a blessing," she now says of the disease. "I'm a different person." She appreciates each day, and her faith in a higher power is only strengthened as she continues to battle. "It saved my life and I'm grateful for that," Kim says. "I never could have done it if the immunotherapy hadn't made it possible."

Appendix A: Types of Immunotherapies Now and Upcoming

1. Summarizing the current state of immunologic science creates an impenetrable list of what's out of date before the ink is dry, and that list is long and growing. It grows each month from research around the world and new clinical data from the thousands of trials currently under way. Speculating on the therapies on the horizon is interesting, but not the goal of this book.
2. Among these are the new bispecifc T cell engagers, or BITE, developed by Amgen. BITE targeted CD19+(positive) B cell malignancies and were approved by the FDA in 2015 under the generic name belimumab; its trade name is Benlysta.
3. Including CD19, CD20, CD33, CD123, HER2, epithelial cell adhesion molecule (EpCAM), BCMA, CEA, and others.
4. Data from the phase 3 clinical trial CheckMate 227, presented at the AACR Annual Meeting in April 2018, showed that among patients with newly diagnosed advanced non-small-cell lung cancer with high-tumor mutational burden, those who received a combination of nivolumab (Opdivo) with ipilimumab (Yervoy) showed significantly improved progression-free survival (PFS) compared with patients who received the previous standard of care chemotherapy.

An American Association for Cancer Research press release cited Memorial Sloan Kettering Cancer Center associate attending physician Dr. Matthew Hellman in reporting that patients who got the combination immunotherapy were 42 percent less likely to show progression of their disease compared with patients who received chemotherapy, a near tripling of progression-free survival at one year (43 percent versus 13 percent, with a minimum follow-up of 11.5 months). The reported objective response rate for patients receiving the combination of checkpoint inhibitors was 45.3 percent, compared with 26.9 percent in those who received the standard of care chemotherapy.

5. Nikolaos Zacharakis et al., "Immune Recognition of Somatic Mutations Leading to Complete Durable Regression in Metastatic Breast Cancer," *Nature Medicine*, 2018, 24:724–730.

6. There are several approaches to making an engineered T cell that both is compatible with the patient's self tissues (and will not attack them as foreign), and also will not be itself attacked as non-self by the patient's own immune system. Some use T cells taken from the cancer patient and bespoke-engineered against their specific cancer; others use a range of donated T cells to create a menu of off-the-shelf therapies compatible with different immune types (MHCs). A promising third route by Dr. Sadelain and others seeks to start from scratch by creating a "universal donor" T cell that can then be retrofitted to recognize whatever tumor antigens you choose. Advances in inserting genes into T cells, greatly enhanced by the advent of CRISPR technology, may allow for the construction of third-generation CAR-T cells, made in a culture from stem cells and capable of recognizing multiple targets, minimizing the toxicity of excessive cytokine release, perhaps even CAR-T cells engineered (or, more precisely, genetically edited) so that they are not susceptible to any of cancer's tricks or down-regulation or exhaustion by factors in the tumor microenvironment.

7. Work in this field is being led by the lab of Dr. Lisa Butterfield at the University of Pittsburgh Department of Medicine and Bernie Fox at the University of Portland.

8. As an example, the target OX-40 was one of the most talked about when I started working on this book; now it's not looking very promising. Another indoleamine 2,3-dioxygenase (IDO) breaks down a fuel (tryptophan) T cells need to proliferate and react. Preliminary combination study data has been confounding.

Index

Memorial Sloan Kettering
  research (1970s), 237n37
side effects of, 139–140
interleukin (IL)
  interleukin-2 (IL-2) and
    Macmillan case, 149, 162
  interleukin-6 (IL-6), 173–175,
    280n16
  interleukin-15 (IL-15), 203
  Memorial Sloan Kettering
    research (1970s), 237n37
  Rosenberg's work with
    interleukin-2 (IL-2), 77–91
  Wolchok's work with
    interleukin-2 (IL-2),
    266–268n47
ipilimumab (Ipi), 132–136, 180–182,
  256n38, 269n58, 287n4. *see also*
  CTLA-4
Isaacs, Alick, 80, 271n1
"It's Time" (Imagine Dragons),
  184–187

J
Jackson Laboratory, 261n16
Jaffee, Elizabeth, 278–279n9
Jenkins, Mark, 252n22
Jenner, Edward, 211–212, 214,
  240n3
Jensen, Michael, 277n4
Johns Hopkins University, 25
Journal of Immunology, on Allison's
  research, 97
*Journal of the American Medical
  Association*
  on Coley's research, 238n39
  on Rosenberg's work, 244–245n35
June, Carl, 169–175, 253n26,
  279n10
June, Cynthia, 170–171
Juno Therapeutics, 176, 277n4

K
Kappler, John, 97–98
Karpaulis, Linda (Taylor), 83–84,
  243n25
Ke$ha, 11–13, 16
Keytruda, 182, 281n2, 285–286n8
kidney cancer
  interleukin-2 (IL-2) approval
    for, 87
  Schwartz's diagnosis, 10, 14–16
  (*see also* Schwartz, Jeff)
killer T cells, 67
Kite Pharma, 176, 281n21
Koch, Robert, 48–49, 235n16
Korman, Alan, 263n33
Krummel, Matthew "Max,"
  101–103, 253n29, 256–257n1,
  256n39
Kymriah, 176–178

L
Lansing, Peter, 253n26
Lawson, David, 142
Leach, Dana, 104, 256n39
Ledbetter, Jeffrey, 101–102,
  252–253nn25–26
leukemia
  acute lymphoblastic leukemia
    (ALL), 170–178
  adult T cell lymphoma-leukemia
    (ATLL), 282n3
  Allison's research on, 93,
    247nn4–5
  defined, 172
  T cell research and, 77
leukocytes (white blood cells)
  discovery of, 60–61
  lymphocytes and, 66
ligands, 209
Lindenmann, Jean, 80,
  270–271n1

About the Author

CHARLES GRAEBER is an award-winning journalist and the *New York Times* bestselling author of *The Good Nurse*. His writing has been anthologized in *The Best American Science Writing*, *The Best American Crime Reporting*, *The Best of 10 Years of National Geographic Adventure*, *The Best of 20 years of WIRED*, and *The Best American Magazine Writing* selected by ASME and the Columbia Journalism School. He is an Alfred P. Sloan fellow and recipient of the Overseas Press Club's Ed Cunningham Award for outstanding international journalism. He currently serves on the board of The Writers Room in New York and Building Markets, an international nonprofit that works to empower refugees from war, poverty, and persecution. Born in Iowa, he is now based in Brooklyn.

Find out more at www.charlesgraeber.com.